21世纪高等教育计算机规划教材

SSM
开发实战教程
Spring+Spring MVC+MyBatis

李西明 陈立为 主编

人民邮电出版社
北京

图书在版编目（CIP）数据

SSM开发实战教程：Spring+Spring MVC+MyBatis / 李西明, 陈立为主编. -- 北京：人民邮电出版社, 2019.7（2024.1重印）
21世纪高等教育计算机规划教材
ISBN 978-7-115-50633-7

Ⅰ. ①S… Ⅱ. ①李… ②陈… Ⅲ. ①JAVA语言—程序设计—高等学校—教材 Ⅳ. ①TP312.8

中国版本图书馆CIP数据核字（2019）第016052号

内 容 提 要

本书详细讲解了当前Java EE开发流行的SSM框架，重点讲述MyBatis、Spring、Spring MVC三大框架的知识、原理与实际应用，以及Spring+Spring MVC+MyBatis三大框架的整合应用。

全书共14章，第1~3章是MyBatis部分，第4~8章为Spring部分，第9~12章为Spring MVC部分，第13章为SSM三大框架的整合，第14章是实战项目。全书共提供60多个丰富的项目案例，将理论知识融合到项目案例中，使读者能更加容易地理解SSM框架的关键技术。本书所有项目案例及实战项目均能直接上机运行，并提供全部源码、课件、习题参考答案等配套资源。

本书可作为高等院校计算机相关专业Java类课程的教材及社会培训机构教材，也适合Java技术爱好者学习或参考。

◆ 主　编　李西明　陈立为
　责任编辑　张　斌
　责任印制　陈　犇
◆ 人民邮电出版社出版发行　北京市丰台区成寿寺路11号
　邮编　100164　电子邮件　315@ptpress.com.cn
　网址　http://www.ptpress.com.cn
　固安县铭成印刷有限公司印刷
◆ 开本：787×1092　1/16
　印张：19.5　　　　　　　　　　　2019年7月第1版
　字数：485千字　　　　　　　　　2024年1月河北第8次印刷

定价：59.80元

读者服务热线：(010)81055256　印装质量热线：(010)81055316
反盗版热线：(010)81055315
广告经营许可证：京东市监广登字20170147号

前言 FOREWORD

学习 SSM 框架的必要性

目前 Java EE 的轻量级开发主要有两种框架组合方式：SSH 与 SSM。SSH 是 Spring、Struts2、Hibernate 三者的组合，相对来说响应会稍微慢一些，学习难度较大，学习周期较长，早期的开发多采用这种方式。随着互联网的发展，开发人员需要更快的响应速度和更快的开发周期，SSM 框架便出现了，它是 Spring、Spring MVC、MyBatis 三大框架的组合。Spring MVC 作为控制层的框架取代 Struts2，能更好地与 Spring 集成，响应速度更快。MyBatis 是半自动化的 ORM 方式，直接使用原生 SQL 语句，无须再用面向对象的 HQL 语句，简单方便。当前的互联网开发采用 SSM 框架越来越多，此外，SSM 框架也是学习 Spring 系列开发技术（如 Spring Boot 和 Spring Cloud 等）的基础。因此，要学习 Java EE，SSM 框架是必须掌握的技术。

本书特色

本书内容丰富全面，实用技术多，提供大量的项目案例，项目案例之间具有很好的连贯性，知识体系由浅入深，层层推进，知识点的讲解不单纯是理论的堆砌，而是以案例为驱动，将知识点融入案例中，读者学习起来更加轻松，更易理解与掌握。

1. 内容丰富

本书涵盖 MyBatis、Spring、Spring MVC 相关的知识，以及它们之间的整合和综合案例。

2. 实践性强

本书把理论知识融合到项目案例中，主要知识点均对应有项目案例，每个案例都有详细的步骤，读者跟着书一步步完成项目案例的同时就掌握了各个知识点。全书提供 60 多个案例，案例都能独立运行，全部提供源码与数据库文件。

项目案例清单

知识章节	案例项目名称	功能
MyBatis	mybatis11、mybatis12 等，15 个完整独立项目	实现 MyBatis 的项目搭建，基本增删改查到动态查询、模糊查询、一对多、多对多、自连接等复杂查询功能，以及 MyBatis 逆向工程
Spring	spring1~spring16，16 个完整独立项目	实现 IOC、AOP、数据库操作、事务管理等功能
Spring MVC	springmvc1~springmvc33 及 springmvc71 等，34 个完整独立项目	实现 Spring MVC 基本框架搭建、参数传递、注解、Ajax、转发与重定向、拦截器、数据验证、类型转换、异常处理、文件上传与下载等
SSM 三大框架整合	springmybatis1、springmybatis2、ssm1，3 个完整独立项目	完成 Spring 对 MyBatis 的整合、事务管理；整合三大框架，实现学生信息管理中的增删改查基本功能
项目实战	综合案例，1 个完整独立项目	实现用户登录、在线浏览与购物车、下订单、结算等流程

3. 连贯性

本书采用案例贯穿的方式讲解，上一个知识点用了一个案例，下一个知识点可以把上一个案例作为素材，继续完善下去，就像搭积木一样，由浅入深，这样学生易于理解，教师也方便备课。本书提供的案例源码，全部运行测试通过。

4. 综合性强

本书最后一章提供了实战项目，把讲解的重点知识贯穿起来，达到很好的巩固与应用的目的。

教学学时建议

内容	学时数	
	理论+实践	实践
MyBatis	16	
Spring	12	
Spring MVC	20	
SSM 三大框架整合	8	
SSM 项目实战		16
合计	72 学时（可根据实际情况适当调整）	

配套服务与支持

本书提供全部案例源码、数据库文件、实战项目源码、教学用 PPT、习题参考答案等，读者可登录人邮教育社区（www.ryjiaoyu.com）下载。

致谢

广州砺锋信息科技有限公司作为多年从事 Java 开发的专业公司为本书的编写提供了大力支持，公司总经理林丽静给本书提出了很多宝贵的意见，提供了很多技术资料，也进行了大量的前期准备工作，在此表示诚挚的感谢。

意见与反馈

由于本书的内容较多，编写时间紧、难度大，而且技术发展也日新月异，书中难免会有不足之处，欢迎读者指正。读者可加入 QQ 群 1108179147 交流互动。

编　者

2019 年 2 月于广州

目录 CONTENTS

第1章 MyBatis 入门 1
1.1 SSM 框架简介 2
1.2 MyBatis 概述 2
1.2.1 ORM 框架原理 2
1.2.2 MyBatis 与 Hibernate 的比较 2
1.2.3 搭建 MyBatis 开发环境 3
1.3 第一个 MyBatis 项目 3
1.4 MyBatis 的工作流程 8
1.5 使用工具类简化第一个项目 9
1.6 利用属性文件读取数据库连接信息 10
1.7 主配置文件简介 11
1.7.1 <setting>标签 12
1.7.2 <typeAliases>标签 12
1.7.3 <typeHandlers>标签 15
1.7.4 <environments>标签 16
1.7.5 <mappers>标签 17
1.8 连接其他数据库 18
上机练习 19
思考题 19

第2章 单表的增删改查 20
2.1 结果映射 resultMap 21
2.2 使用 selectOne 方法查询单条记录 23
2.3 使用 insert 方法添加记录 24
2.3.1 主键非自增长 24
2.3.2 主键值由数据库自增长 26
2.4 使用 delete 方法删除记录 29
2.5 使用 update 方法修改记录 30
2.6 模糊查询 31
2.7 动态查询 33
2.7.1 <if>标签 33
2.7.2 <where/>标签 35
2.7.3 使用 Map 封装查询条件 36
2.7.4 <choose/>标签 38
2.7.5 使用<foreach/>标签遍历数组 40
2.7.6 使用<foreach/>标签遍历泛型为基本类型的 List 41
2.7.7 使用<foreach/>标签遍历泛型为自定义类型的 List 42
2.7.8 <sql/>标签 43
2.8 分页查询基础 44
2.9 getMapper 面向接口编程 46
2.10 多参数查询 47
2.11 MyBatis 读写 Oracle 大对象数据类型 48
2.12 MyBatis 调用存储过程 53
2.13 MyBatis 逆向工程 54
上机练习 56
思考题 57

第3章 多表关联查询 58
3.1 一对多查询 59
3.2 多对一关联查询 63
3.3 自连接 65
3.3.1 使用多对一的方式实现自连接 65
3.3.2 使用一对多方式实现自连接 68
3.4 多对多查询 70
上机练习 74
思考题 75

第4章 Spring 入门 76
4.1 Spring 概述 77
4.1.1 Spring 的体系结构 77
4.1.2 Spring 的开发环境 78
4.2 第一个 Spring 程序 78

上机练习 …………………………………… 81
思考题 ……………………………………… 81

第 5 章　Spring 控制反转 …………… 82

5.1　依赖注入 ……………………………… 83
5.2　Spring 配置文件中 Bean
　　 的属性 ……………………………… 85
5.3　Bean 的作用域 ………………………… 86
5.4　基于 XML 的依赖注入 ……………… 87
　　 5.4.1　设值注入 ……………………… 87
　　 5.4.2　构造注入 ……………………… 89
　　 5.4.3　p 命名空间注入 ……………… 90
　　 5.4.4　各种数据类型的注入 ………… 90
5.5　自动注入 ……………………………… 92
　　 5.5.1　byName 方式自动注入 ……… 92
　　 5.5.2　byType 方式自动注入 ……… 94
5.6　Spring 配置文件的拆分 ……………… 94
　　 5.6.1　拆分为若干个平等关系的
　　　　　 配置文件 ………………………… 95
　　 5.6.2　拆分为父子关系的若干个
　　　　　 配置文件 ………………………… 95
5.7　基于注解的依赖注入 ………………… 95
　　 5.7.1　使用注解@Component
　　　　　 定义 Bean ……………………… 96
　　 5.7.2　Bean 的作用域@Scope ……… 97
　　 5.7.3　基本类型属性注入@Value … 97
　　 5.7.4　按类型注入域属性@Autowired … 98
　　 5.7.5　按名称注入域属性
　　　　　 @Autowired 与@Qualifier … 99
　　 5.7.6　域属性注解@Resource ……… 100
　　 5.7.7　XML 配置方式与注解方式
　　　　　 的比较 …………………………… 100
上机练习 …………………………………… 100
思考题 ……………………………………… 101

第 6 章　Spring 面向切面编程 ……… 102

6.1　传统编程模式的弊端 ………………… 103
6.2　AOP 初试身手 ……………………… 105
6.3　AspectJ ………………………………… 108
　　 6.3.1　异常通知 ……………………… 109

　　 6.3.2　环绕通知 ……………………… 110
6.4　使用注解实现通知 …………………… 112
6.5　使用 XML 定义切面 ………………… 113
　　 6.5.1　切面不获取切点参数 ………… 114
　　 6.5.2　切面获取切点方法的参数
　　　　　 与返回值 ………………………… 115
上机练习 …………………………………… 118
思考题 ……………………………………… 118

第 7 章　Spring 操作数据库 ………… 119

7.1　JdbcTemplate 数据源 ………………… 120
　　 7.1.1　DriverManagerDataSource
　　　　　 数据源 …………………………… 120
　　 7.1.2　DBCP 数据源
　　　　　 BasicDataSource ……………… 121
　　 7.1.3　C3P0 数据源
　　　　　 ComboPooledDataSource …… 122
　　 7.1.4　使用属性文件读取数据库
　　　　　 连接信息 ………………………… 122
7.2　JdbcTemplate 方法的应用 …………… 123
上机练习 …………………………………… 128
思考题 ……………………………………… 128

第 8 章　Spring 事务管理 …………… 129

8.1　Spring 事务管理接口 ………………… 130
　　 8.1.1　事务管理器接口
　　　　　 PlatformTransactionManager … 130
　　 8.1.2　事务定义接口
　　　　　 TransactionDefinition ………… 130
8.2　Spring 事务管理的实现方法 ………… 132
　　 8.2.1　没有事务管理的情况分析 …… 132
　　 8.2.2　通过配置 XML 实现事务
　　　　　 管理 ……………………………… 134
　　 8.2.3　利用注解实现事务管理 ……… 137
　　 8.2.4　在业务层实现事务管理 ……… 139
上机练习 …………………………………… 141
思考题 ……………………………………… 141

第 9 章　Spring MVC 入门 …………… 142

9.1　Spring MVC 简介 …………………… 143

9.1.1　Spring MVC 的优点·················143
9.1.2　SpringMVC 的运行原理··········143
9.2　第一个 Spring MVC 程序··················144
9.2.1　开发环境·······························144
9.2.2　第一个 Spring MVC 程序······145
上机练习···151
思考题··151

第 10 章　Spring MVC 注解式开发·········152

10.1　第一个注解式开发程序·················153
10.2　核心控制器 DispatcherServlet 的配置··155
10.3　@Controller 注解··························156
10.4　@RequestMapping 注解················156
　　10.4.1　注解用于方法上··················156
　　10.4.2　注解用于类上·····················158
　　10.4.3　请求的提交方式·················161
　　10.4.4　请求 URI 中使用通配符····162
　　10.4.5　请求中携带参数·················163
10.5　客户端到处理器的参数传递·········164
　　10.5.1　基本类型做形式参数········164
　　10.5.2　中文乱码问题·····················166
　　10.5.3　实体 Bean 做形参·············166
　　10.5.4　实体 Bean 含对象属性······167
　　10.5.5　路径变量····························168
　　10.5.6　RESTful 风格编程·············169
　　10.5.7　HttpServletRequest 参数···172
　　10.5.8　接收数组类型的请求参数··173
10.6　服务端到客户端的参数传递·········173
10.7　控制器方法返回 String 类型·······174
　　10.7.1　返回 View 对象名···············174
　　10.7.2　使用 Model 参数················175
　　10.7.3　使用 HttpSerlvetRequest 参数·······································176
　　10.7.4　使用 HttpSession 参数······176
10.8　控制器方法返回 void 类型··········176
　　10.8.1　使用 ServletAPI 参数········177
　　10.8.2　Ajax 响应···························177

10.9　控制器方法返回 Object 类型······179
10.10　Ajax/JSON 专项突破···················183
　　10.10.1　服务端接收对象返回 JSON 字符串································183
　　10.10.2　服务端接收 Bean 返回 JSON 对象·······························184
　　10.10.3　服务端接收属性返回 JSON 对象·······························185
　　10.10.4　客户端发送 JSON 字符串 返回 JSON 对象···················186
　　10.10.5　数据接收与返回的格式 限制·······································187
　　10.10.6　直接输出响应字符串······188
上机练习···189
思考题··190

第 11 章　Spring MVC 关键技术·············191

11.1　转发与重定向································192
　　11.1.1　请求转发到其他页面········192
　　11.1.2　请求转发到其他控制器····194
　　11.1.3　返回 String 时的请求转发··194
　　11.1.4　请求重定向到其他页面····195
　　11.1.5　请求重定向到其他控制器··197
　　11.1.6　返回 String 时的重定向···198
　　11.1.7　返回 void 时的请求转发···200
　　11.1.8　返回 void 时的重定向······200
11.2　异常处理·······································201
　　11.2.1　SimpleMappingException Resolver 异常处理器·············201
　　11.2.2　HandlerExceptionResolver 接口处理异常·······················204
　　11.2.3　使用@ExceptionHandler 注解实现异常处理················205
11.3　类型转换器···································208
　　11.3.1　自定义类型转换器 Converter·····························209
　　11.3.2　接收多种格式的日期 类型转换····························211

3

11.3.3　类型转换发生异常后的
　　　　　　数据回显 …………………… 212
　　　11.3.4　简化类型转换发生异常后
　　　　　　的提示信息 ………………… 215
　11.4　数据验证 ……………………………… 217
　11.5　文件上传 ……………………………… 221
　　　11.5.1　上传单个文件 ………………… 221
　　　11.5.2　上传多个文件 ………………… 224
　11.6　文件下载 ……………………………… 226
　11.7　拦截器 ………………………………… 228
　　　11.7.1　单个拦截器的执行流程 ……… 228
　　　11.7.2　多个拦截器的执行 …………… 230
　　　11.7.3　权限拦截器 …………………… 232
　11.8　静态资源访问 ………………………… 234
　　　11.8.1　使用 Tomcat 中名为 default
　　　　　　的 Servlet ………………………… 235
　　　11.8.2　使用<mvc:default-servlet-
　　　　　　handler/> ………………………… 235
　　　11.8.3　使用<mvc:resources/> ……… 236
　上机练习 ……………………………………… 237
　思考题 ………………………………………… 237

第 12 章　Spring MVC
　　　　　　表单标签 ………………… 238

　12.1　表单标签 ……………………………… 239
　　　12.1.1　form 标签 ……………………… 239
　　　12.1.2　input 标签 ……………………… 240
　　　12.1.3　password 标签 ………………… 240
　　　12.1.4　checkbox 标签 ………………… 240
　　　12.1.5　checkboxes 标签 ……………… 241
　　　12.1.6　radiobutton 与 radiobuttons
　　　　　　标签 ……………………………… 242
　　　12.1.7　select 与 option/options 标签 …… 243
　12.2　表单标签使用综合案例 ……………… 244
　上机练习 ……………………………………… 250
　思考题 ………………………………………… 250

第 13 章　SSM 三大框架整合 …… 251

　13.1　Spring 整合 MyBatis ………………… 252
　　　13.1.1　Spring 整合 MyBatis
　　　　　　开发环境 ………………………… 252
　　　13.1.2　DAO 接口实现类开发整合 …… 252
　　　13.1.3　DAO 接口无实现类
　　　　　　开发整合 ………………………… 259
　13.2　SSM 整合案例 ………………………… 261
　上机练习 ……………………………………… 273
　思考题 ………………………………………… 273

第 14 章　SSM 项目实战 …………… 274

　14.1　项目需求分析 ………………………… 275
　14.2　搭建 SSM 框架 ……………………… 275
　14.3　首页与用户登录模块设计 …………… 277
　14.4　商品查询与分页模块设计 …………… 283
　14.5　商品详情模块设计 …………………… 288
　14.6　购物车模块设计 ……………………… 290
　14.7　订单处理与模拟结算模块设计 ……… 297
　上机练习 ……………………………………… 302
　思考题 ………………………………………… 302

参考文献 ……………………………………… 303

第 1 章　MyBatis 入门

本章目标

- 掌握 MyBatis 原理与工作流程
- 学会搭建 MyBatis 开发环境
- 掌握 MyBatis 配置文件的方法
- 掌握 MyBatis 映射文件的方法
- 学会使用工具类简化开发流程
- 学会使用属性文件配置数据库连接

1.1 SSM 框架简介

SSM 框架是 Spring、Spring MVC 和 MyBatis 三大框架的组合,是目前主流的 Java EE 企业级框架,适用于搭建各种大型的企业级应用系统。SSM 采用标准的 MVC 模式,将整个系统划分为数据访问层(DAO 层)、业务辑层、控制层、表示层,使用 MyBatis 管理 DAO 层,作为对象数据的持久化引擎,使用 Spring MVC 进行请求转发与视图管理,使用 Spring 实现业务对象管理并整合其他框架。

1.2 MyBatis 概述

MyBatis 是当前 Java Web 开发中流行的持久化 ORM 框架,它对 JDBC 进行了封装与简化,无须 JDBC 的注册驱动、创建 Connection 连接、配置 Statement 等烦琐过程,大大减少了 JDBC 代码,使开发者只需专注于 SQL 语句设计即可。MyBatis 通过内部机制将 Java 类(对象)持久化为数据库表中的记录,反之也可将数据库中的记录转化为 Java 类(对象)。在三层架构开发中,MyBatis 作用在数据访问层,它让数据访问层的开发变得简单、高效。

1.2.1 ORM 框架原理

Java 程序常常要连接并操作数据库,但两者的数据类型往往并不匹配,Java 是面向对象的语言,Java 语言中操作的单元是类与对象,而数据库的数据格式是关系类型。为了匹配 Java 面向对象与关系数据库的数据类型,人们发明了 ORM 框架(Object Relational Mapping,对象关系映射),用于将 Java 中的对象映射成数据库中的记录,对象中的属性映射为数据库表中的字段,程序员可使用面向对象的编程方式来操作数据库。例如,在程序代码中添加一个对象,则数据库中相应添加一行记录,在程序代码中删除一个对象,则数据库中相应删除了一条记录。比较常见的 ORM 框架有 Hibernate、MyBatis。

ORM 框架的原理如图 1.1 所示。其中,CRUD 是在做计算处理时的增加(Create)、读取查询(Retrieve)、更新(Update)和删除(Delete)4 个单词的首字母缩写,POJO(Plain Ordinary Java Object)为简单的 Java 对象,实际就是普通 JavaBeans。

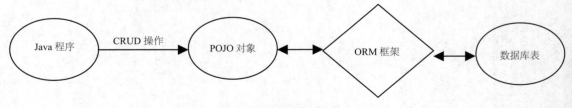

图 1.1 ORM 框架原理图

1.2.2 MyBatis 与 Hibernate 的比较

Hibernate 框架是"全自动"的 ORM,使用完全的数据库封装机制实现对象和数据库表之间的映射,自动生成与执行 SQL 语句,数据库无关性比较好,方便移植。

MyBatis 框架是"半自动"的 ORM,需要程序员自行编写 SQL 语句,MyBatis 再通过映射文件将返回的结果字段映射到指定对象。MyBatis 无法做到数据库无关性,每次只能针对一个数据库做开发。

与 Hibernate 相比,MyBatis 具有以下几个优点。

① SQL 语句与代码分离。只需要在 XML 映射文件中配置 SQL 语句,而无须在 Java 代码中编写 SQL 语句,给程序的维护带来了很大便利。

② 手写 SQL 语句,灵活方便,查询效率高,能够完成较复杂的查询。

③ 采用原生 SQL 语言,无须学习面向对象的 HQL 语言,相对简单易学,容易上手。

1.2.3 搭建 MyBatis 开发环境

在项目中使用 MyBatis 的基本步骤导入相关 JAR 包、编写配置文件和映射文件、创建接口、创建实体类,最后是设计 DAO 层的实现类。

本书使用的 MyBatis 版本为 MyBatis 3.4.5,可在 GitHub 网站下载,下载 MyBatis-3.4.5.zip 解压后的文件结构如图 1.2 所示。

图 1.2 MyBatis-3.4.5.zip 解压后的文件结构

上述解压目录中,lib 子目录放置支持 MyBatis 运行的多个 JAR 包,mybatis-3.4.5.jar 是 MyBatis 的核心 JAR 包,必须导入项目中。此外,还需要导入连接 MySQL 数据库的驱动包 mysql-connector-java-5.1.37.jar,所以 MySQL 项目至少要导入上述两个 JAR 包,具体操作过程见 1.3 节。本书 Java 开发工具采用 Eclipse 4.7,JDK 版本为 1.8,数据库采用 MySQL 5.7。

1.3 第一个 MyBatis 项目

项目案例:新建项目 mybatis11,查询 MySQL 数据库 studentdb 中的表 student 中的所有学生信息,在控制台输出。(项目源码参见本书配套资源:第 1 章/第 1 个 mybatis 项目/mybatis11)

实现步骤：

（1）在 MySQL 中创建数据库 studentdb，创建表 student，并添加若干测试用数据记录，SQL 语句如下：

```
CREATE DATABASE studentdb;
USE studentdb ;
DROP TABLE IF EXISTS student ;
CREATE TABLE student (
    id int(11) NOT NULL,
    studentname varchar(20) DEFAULT NULL,
    gender char(2) DEFAULT NULL,
    age int(11) DEFAULT NULL,
    PRIMARY KEY ( id )
)
insert into student ( id , studentname , gender , age ) values (1,'张飞','男',18),(2,'李白','男',20),(3,'张无忌','男',19),(4,'赵敏','女',17);
```

（2）在 Eclipse 中新建 Web 项目 mybatis11，将下载的 MyBatis 核心 JAR 包以及 MySQL 数据库驱动 JAR 包导入（复制）到目录 WebContent/WEB-INF/lib 下。完成后的 lib 下的 JAR 包清单如图 1.3 所示。

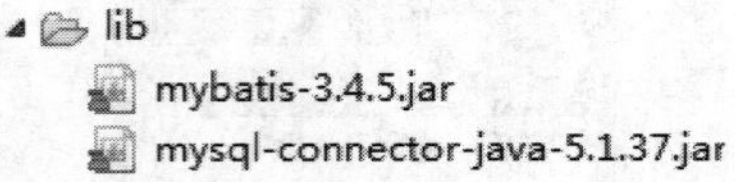

图 1.3　搭建 MyBatis 项目所需 JAR 包清单

【**注意**】将 JAR 包复制到 lib 下并不意味着大功告成，还需选中全部 JAR 包，再单击鼠标右键在弹出的菜单中选择 Build Path，选择 Add to Build Path 添加到构建路径。

（3）为了方便查看控制台输出 SQL 语句，还要配置 log4j，在项目目录 src 下创建 log4j.properties 文件，输入内容如下：

```
# Global logging configuration
log4j.rootLogger=ERROR, stdout
# MyBatis logging configuration...
log4j.logger.com.lifeng=DEBUG
# Console output...
log4j.appender.stdout=org.apache.log4j.ConsoleAppender
log4j.appender.stdout.layout=org.apache.log4j.PatternLayout
log4j.appender.stdout.layout.ConversionPattern=%5p [%t] - %m%n
```

其中 log4j.logger.com.lifeng=DEBUG 里面的 com.lifeng 是包名，根据项目不同包名也会不同。本项目的类放在包 com.lifeng 下。DEBUG 是日志记录的一种级别，表示可以显示所执行的 SQL 语句、参数值、对数据库的影响等信息，若改为 TRACE 级别，则还可进一步显示查询出的每条记录的每个字段名及值。

以上内容无须死记，可在下载的 mybatis-3.4.5 文件夹里的 mybatis-3.4.5.pdf 文件的 Logging 节里找到模板，再简单修改即可。

（4）在项目 src 目录下新建包 com.lifeng.entity，创建实体类 Student 如下：

```
package com.lifeng.entity;
public class Student {
    private String sid;
    private String sname;
    private String sex;
    private int age;
    public Student(){}
    public Student(String sid,String sname,String sex,int age){
        this.sid=sid;
        his.sname=sname;
        his.sex=sex;
        his.age=age;
    }
    public void show(){
        System.out.println("学生编号:"+sid+" 学生姓名:"+sname+" 学生性别:"+sex+" 学生年龄:"+age);
    }
    //省略 setter,getter 方法
}
```

（5）新建包 com.lifeng.dao，新建 DAO 层接口 IStudentDao，代码如下所示：

```
public interface IStudentDao {
    public List<Student> findAllStudents();//查找全部学生
}
```

（6）在包 com.lifeng.dao 下创建映射文件 StudentMapper.xml，代码及其解释如下：

```
<?xml version="1.0" encoding="utf-8" ?>
<!DOCTYPE mapper PUBLIC "-//mybatis.org//DTD Mapper 3.0//EN"
"http://mybatis.org/dtd/mybatis-3-mapper.dtd">
<mapper namespace="com.lifeng.dao.IStudentDao">
    <select id="findAllStudents" resultType="com.lifeng.entity.Student">
        SELECT
            id as sid,
            studentname as sname,
            gender as sex,
            age
        FROM STUDENT
    </select>
</mapper>
```

（注释：命名空间；SQL 语句；SQL 语句查询结果映射为 Student 对象类型）

映射文件主要实现 SQL 语句与 Java 对象之间的映射，使 SQL 语句查询出来的关系型数据能封装成 Java 对象。映射文件通常用 POJO+Mapper 命名，一般在 DAO 层中与接口放在一起。这里 POJO 为 Student，所以映射文件名称定为 StudentMapper.xml。

映射文件中，<!DOCTYPE mapper PUBLIC "-//mybatis.org//DTD Mapper 3.0//EN" "http://mybatis.org/dtd/mybatis-3-mapper.dtd">是映射文件的约束信息，可以从 MyBatis 解压文件中的 mybatis-3.4.5.pdf 文档中找到，在该文档中搜索 "mybatis-3-mapper.dtd" 关键字，即可找到映射文件的约束。

图 1.4 所示是从该 PDF 文档中搜索到的，直接复制过来即可。

```
<?xml version="1.0" encoding="UTF-8" ?>
<!DOCTYPE mapper
  PUBLIC "-//mybatis.org//DTD Mapper 3.0//EN"
  "http://mybatis.org/dtd/mybatis-3-mapper.dtd">
<mapper namespace="org.mybatis.example.BlogMapper">
  <select id="selectBlog" resultType="Blog">
    select * from Blog where id = #{id}
  </select>
</mapper>
```

约束信息

图 1.4　映射文件的约束信息模板

映射文件中的一级标签<mapper></mapper>里面可包含多对<select>标签或者<insert><delete><update>等标签。

<mapper>标签的 namepace 属性，用于标识映射文件，通常其值设置成对应接口的全路径名称。<select></select>标签用于设计 SQL 语句，其标签内部只能是 select 查询语句，同理<insert></insert>标签内部只能是 insert 语句，<update></update>标签内部只能是 update 语句，<delete></delete>标签内部只能是 delete 语句。

<select><insert><delete><update>等标签的 id 属性用于唯一标识该 SQL 语句块，Java 代码使用该标识来找到对应的 SQL 语句块。其值设置成与接口中对应的方法名称一致。

（7）在 src 下新建 XML 文件 mybatis-config.xml 作为主配置文件，完整代码如下：

```
<?xml version="1.0" encoding="UTF-8"?>
<!DOCTYPE configuration PUBLIC "-//mybatis.org//DTD Config 3.0//EN"
"http://mybatis.org/dtd/mybatis-3-config.dtd">
<configuration>
    <!-- 配置环境,默认环境id为development -->
    <environments default="development">
        <environment id="development">
        <!-- 配置事务管理类型为 JDBC -->
            <transactionManager type="JDBC"/>
            <!-- 配置数据源类型为连接池 -->
            <dataSource type="POOLED">
            <!-- 分别配置数据库连接的驱动,url,用户名,密码 -->
                <property name="driver" value="com.mysql.jdbc.Driver"/>
                <propertyname="url"value="jdbc:mysql://localhost:3306/studentdb?characterEncoding=utf8"/>
                <property name="username" value="root"/>
                <property name="password" value="root"/>
            </dataSource>
        </environment>
    </environments>
    <!-- 配置映射文件的位置,可以有多个映射文件 -->
    <mappers>
        <mapper resource="com/lifeng/dao/StudentMapper.xml"/>
    </mappers>
</configuration>
```

上边的代码看起来很多，但无须死记，相关的模板可以从 MyBatis 下载文件解压目录下的 mybatis-3.4.5.pdf 文档中找到，在该文档中搜索 "mybatis-3-config.dtd" 关键字，即可找到映射文件的模板，如图 1.5 所示。上面代码第 2~3 行为主配置文件约束信息，提供有模板，其他部分如数据库连接的配置同样可参考文档 mybatis-3.4.5.pdf 中的模板。

```xml
<?xml version="1.0" encoding="UTF-8" ?>
<!DOCTYPE configuration
  PUBLIC "-//mybatis.org//DTD Config 3.0//EN"
  "http://mybatis.org/dtd/mybatis-3-config.dtd">
<configuration>
  <environments default="development">
    <environment id="development">
      <transactionManager type="JDBC"/>
      <dataSource type="POOLED">
        <property name="driver" value="${driver}"/>
        <property name="url" value="${url}"/>
        <property name="username" value="${username}"/>
        <property name="password" value="${password}"/>
      </dataSource>
    </environment>
  </environments>
  <mappers>
    <mapper resource="org/mybatis/example/BlogMapper.xml"/>
  </mappers>
</configuration>
```

约束信息

图 1.5　主配置文件模板

使用时只需把整个模板的内容复制过来再稍加修改就可以了。大部分内容不用改动，这里只修改了数据库连接的 4 条配置，给出了具体的数据库连接信息，以替换模板中的 ${} 占位符。

```xml
<property name="driver" value="com.mysql.jdbc.Driver"/>
<property name="url" value="jdbc:mysql://localhost:3306/studentdb?characterEncoding=utf8"/>
<property name="username" value="root"/>
<property name="password" value="root"/>
```

此外还修改了 <mapper> 节点，用于指出映射文件的全路径名（位置）。

```xml
<mapper resource="com/lifeng/dao/StudentMapper.xml"/>
```

【注意】如果一个项目有多个映射文件，则主配置文件需要多个 <mapper> 节点。这里的包路径用到符号 "/"，而不是 "."。主配置文件名可以随意命名，但通常命名为 mybatis-config.xml。

（8）创建 DAO 层实现类 StudentDaoImpl.java，代码如下：

```java
public class StudentDaoImpl implements IStudentDao{
    @Override
    public List<Student> findAllStudents() {
        SqlSession session = null;
        List<Student> list = new ArrayList<Student>();
        try {
            //1.读取主配置文件 mybatis-config.xml
            String resource = "mybatis-config.xml";
            Reader reader = Resources.getResourceAsReader(resource);
            //2.根据主配置文件 mybatis-config.xml 构建 SqlSessionFactory 对象 factory
            SqlSessionFactoryBuilder builder = new SqlSessionFactoryBuilder();
            SqlSessionFactory factory = builder.build(reader);
            //3.根据 SqlSessionFactory 对象创建 SqlSession 对象 session
            session = factory.openSession();
            //4.调用 SqlSession 对象 session 的 selectList 方法执行查询数据库的操作，返回映射后的结果集合
            list = session.selectList("com.lifeng.dao.IStudentDao.findAllStudents");
        } catch (IOException e1) {
            e1.printStackTrace();
        }
        return list;
    }
}
```

上述代码中的 session.selectList()是 SqlSession 类的方法，用于查询记录集合。

（9）创建测试类 TestStudent1，新建包 com.lifeng.test，包下新建类 TestStudent1，代码如下：

```java
public class TestStudent1 {
    public static void main(String[] args) {
        findAllStudents();
    }
    public static void findAllStudents(){
        IStudentDao studentDao=new StudentDaoImpl();
        List<Student> sList=studentDao.findAllStudents();
        for(int i=0;i<sList.size();i++){
            sList.get(i).show();
        }
    }
}
```

运行测试类 TestStudent1 结果如下：

```
DEBUG [main] - ==>  Preparing: SELECT id as sid, studentname as sname, gender as sex, age FROM STUDENT
DEBUG [main] - ==> Parameters: 
DEBUG [main] - <==      Total: 4
学生编号:1 学生姓名:张飞 学生性别:男 学生年龄:18
学生编号:2 学生姓名:李白 学生性别:男 学生年龄:20
学生编号:3 学生姓名:张无忌 学生性别:男 学生年龄:19
学生编号:4 学生姓名:赵敏 学生性别:女 学生年龄:17
```

可见，数据库中的记录全都读出来了，由于之前配置了 log4j，级别为 DEBUG，所以 SQL 查询语句、参数、受影响记录数也显示出来了，方便程序员查看效果与检查调试。如果将级别改为 TRACE，则结果如下：

```
DEBUG [main] - ==>  Preparing: SELECT id as sid, studentname as sname, gender as sex, age FROM STUDENT
DEBUG [main] - ==> Parameters: 
TRACE [main] - <==    Columns: sid, sname, sex, age
TRACE [main] - <==        Row: 1, 张飞, 男, 18
TRACE [main] - <==        Row: 2, 李白, 男, 20
TRACE [main] - <==        Row: 3, 张无忌, 男, 19
TRACE [main] - <==        Row: 4, 赵敏, 女, 17
DEBUG [main] - <==      Total: 4
学生编号:1 学生姓名:张飞 学生性别:男 学生年龄:18
学生编号:2 学生姓名:李白 学生性别:男 学生年龄:20
学生编号:3 学生姓名:张无忌 学生性别:男 学生年龄:19
学生编号:4 学生姓名:赵敏 学生性别:女 学生年龄:17
```

可见级别改为 TRACE 后，显示的 SQL 更为详细，本项目无须太过详细，改回 DEBUG 即可。

1.4　MyBatis 的工作流程

通过第一个 MyBatis 项目，可以大概了解 MyBatis 的开发流程，理解其工作流程。结合上述项目，MyBatis 的工作流程如下。

（1）读取主配置文件 mybatis-config.xml，获得运行环境和数据库连接。

（2）加载映射文件，如 StudentMapper.xml。
（3）根据主配置文件 mybatis-config.xml 构建会话工厂 SqlSessionFactory 对象。
（4）根据会话工厂 SqlSessionFactory 对象创建 SqlSession 对象，再调用 SqlSession 对象的各种增删改查方法。例如，可调用 selectList()方法查找记录集合。
（5）底层定义的 Executor 接口操作数据库。
（6）底层对输入参数进行映射，在执行 SQL 前将输入的 Java 对象映射到 SQL 中。
（7）底层将输出结果映射为 Java 对象。
最后 3 个步骤无须程序员参与，MyBatis 底层会自动执行。

1.5　使用工具类简化第一个项目

上面的案例实现了数据的查询操作，如果还要实现添加、删除、修改等操作，则实现步骤类似。由于每一次执行 SqlSession 的方法前，均需先读取主配置文件，再创建 SqlSessionFactory 会话工厂，最后创建 SqlSession 对象，这一系列过程比较烦琐，所以可以将获取 SqlSession 对象的一系列程序代码封装定义为一个工具类方法 getSession()，放在单独的一个类 MyBatisUtil 中，将来只要调用这个方法，即可获取 SqlSession 对象，从而简化所有的增删改查操作。

项目案例：使用工具类 MyBatisUtil 简化第 1 个项目的 DAO 层实现类 StudentDaoImpl 的 findAllStudents 方法。（项目源码参见本书配套资源：第 1 章/工具类简化第 1 个 mybatis 项目/mybatis12）

实现步骤：

（1）复制第一个项目 mybatis11 为 mybatis12，新建包 com.lifeng.utils，新建类 MyBatisUtil.java，代码如下。该类提供两个方法，一个用来获取 SqlSession 对象，一个用来关闭 SqlSession 对象。

【注意】项目复制粘贴后还需用鼠标右键单击粘贴后的新项目，然后单击 Properties，选择 Web Project Settings，在 Context root 中更改旧项目名称为新项目名称。

```java
public class MyBatisUtil {
    private MyBatisUtil(){
    }
    private static final String RESOURCE = "mybatis-config.xml";
    private static SqlSessionFactory sqlSessionFactory = null;
    private static ThreadLocal<SqlSession> threadLocal = new ThreadLocal<SqlSession>();
    static {
        Reader reader = null;
        try {
            reader = Resources.getResourceAsReader(RESOURCE);
            SqlSessionFactoryBuilder builder = new SqlSessionFactoryBuilder();
            sqlSessionFactory = builder.build(reader);
        } catch (Exception e1) {
            e1.printStackTrace();
            throw new ExceptionInInitializerError("初始化 MyBatis 失败，请检查配置文件或数据库");
        }
    }
    public static SqlSessionFactory getSqlSessionFactory(){
        return sqlSessionFactory;
    }
}
```

```java
        public static SqlSession getSession(){
            SqlSession session = threadLocal.get();
            // 如果session为null，则打开一个新的session
            if (session == null){
                session = (sqlSessionFactory !=null) ?sqlSessionFactory.openSession():null;
                threadLocal.set(session);
            }
            return session;
        }
        public static void closeSession(){
            SqlSession session = (SqlSession) threadLocal.get();
            threadLocal.set(null);
            if (session !=null){
                session.close();
            }
        }
    }
```

以上代码无须死记，只是工具而已，复制借用过来即可。

【注意】将来使用 Spring 整合 MyBatis 后，这个工具类将不再使用，而是用 Spring 提供的工具代替。

（2）修改接口 IStudentDao 的实现类 StudentDaoImpl，代码如下：

```java
public class StudentDaoImpl implements IStudentDao{
    @Override
    public List<Student> findAllStudents() {
        SqlSession session = null;
        List<Student> list = new ArrayList<Student>();
        try {
            session = MyBatisUtil.getSession();
            list = session.selectList("com.lifeng.dao.IStudentDao.findAllStudents");
        } catch (Exception e) {
            e.printStackTrace();
        }
        return list;
    }
}
```

> 通过工具类获取了 SqlSession 对象

对比原来的 findAllStudents 方法，现在的 findAllStudents 方法的代码简化了不少，其他方面没变化。运行该测试类可得到同样的效果。其他增删改查的方法都可这样做，将大大减少代码冗余。

1.6 利用属性文件读取数据库连接信息

上面案例中的 mybatis-config.xml 主配置文件中，关于数据库的连接信息，分别指定了 Java 连接数据库的 4 大要素：驱动、URL、数据库用户名、密码。

为了方便对数据库连接的管理，可以快速替换到不同的数据库，连接数据库的 4 大要素数据一般都是单独存放在一个专门的属性文件中的，MyBatis 主配置文件再从这个属性文件中读取这些数据。利用属性文件读取数据库连接信息可以方便更换不同的数据库，只需要修改属性文件或者替换为另外一个属性文件即可，可移植性较好。

项目案例：利用属性文件实现 mybaytis 项目的数据库连接。（项目源码参见本书配套资源：第 1

章/利用属性文件实现数据库连接/mybatis13）

实现步骤：

（1）复制上一个项目 mybatis12 为 mybatis13，在 src 下新建一个文件，命名为 jdbc.properties，代码如下：

```
jdbc.driver=com.mysql.jdbc.Driver
jdbc.url=jdbc:mysql://localhost:3306/studentdb
jdbc.username=root
jdbc.password=root
```

上述内容包括了连接数据库的 4 大要素的具体值。

（2）修改主配置文件 mybatis-config.xml，代码如下：

```xml
<?xml version="1.0" encoding="UTF-8"?>
<!DOCTYPE configuration PUBLIC "-//mybatis.org//DTD Config 3.0//EN"
 "http://mybatis.org/dtd/mybatis-3-config.dtd">
<configuration>
    <!-- 注册属性文件 -->
    <properties resource="jdbc.properties"/>
    <environments default="development">
        <environment id="development">
            <transactionManager type="JDBC"/>
            <dataSource type="POOLED">
                <property name="driver" value="${jdbc.driver}"/>
                <property name="url" value="${jdbc.url}"/>
                <property name="username" value="${jdbc.username}"/>
                <property name="password" value="${jdbc.password}"/>
            </dataSource>
        </environment>
    </environments>
    <mappers>
        <mapper resource="com/lifeng/dao/StudentMapper.xml"/>
    </mappers>
</configuration>
```

1. 先在此位置注册属性文件
2. 修改 value 为此种 EL 格式

（3）运行测试，结果不变。将来数据库要更换，只需更改属性文件 jdbc.properties 的有关信息即可，无须改动配置文件，非常实用。

1.7 主配置文件简介

主配置文件 mybatis-config.xml 用来配置系统运行环境，包含事务管理方式、数据库连接类型与信息、指定映射文件等。除了上面第一个项目用到的一些标签之外，还有其他一些标签（标签也可称为节点或元素），MyBatis 主配置文件的所有主要标签如图 1.6 所示。

<configuration>下的所有子标签并不是都必须配置，但若需要配置时，必须按图 1.6 所示的先后顺序来配置，否则 MyBatis 会报错。

<properties>标签的作用是将内部的配置转化为外部的配置，从而能够动态地替换内部定义的属性。例如，数据库的连接信息，原来是在内部配置，通过<properties>属性，让系统读取外部的属性文件，第 1.5 节就是<properties>的典型应用案例。

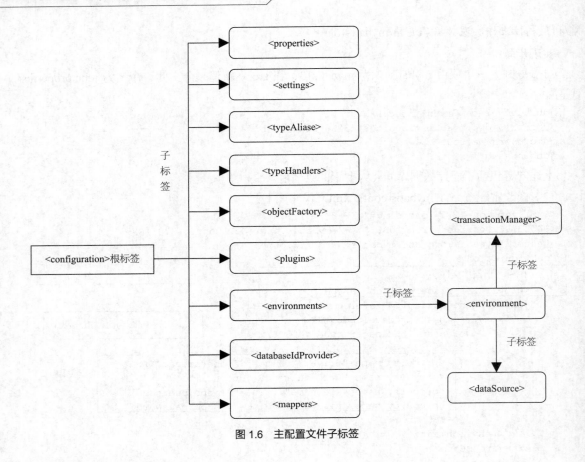

图 1.6 主配置文件子标签

1.7.1 \<setting\>标签

\<setting\>标签用来改变 MyBatis 运行时的行为,如开启延迟加载以及二级缓存。启用延迟加载以及二级缓存有利于提高系统的性能,但在小型系统及硬件性能较强的环境下效果并不明显,一般不配置也可以。

下面代码演示可能用到的一些配置,每一项的作用参见代码注释:

```
<settings>
        <!-- 是否开启缓存 -->
        <setting name="cacheEnabled" value="true"/>
        <!-- 是否开启延迟加载,如果开启的话所有关联对象都会延迟加载 -->
        <setting name="lazyLoadingEnabled" value="true"/>
        <!-- 是否启用关联对象属性的延迟加载,如果启用,对任意延迟属性的调用都会
        使用带有延迟加载属性的对象完整加载,否则每种属性都按需加载 -->
        <setting name="aggressiveLazyLoading" value="true"/>
</settings>
```

1.7.2 \<typeAliases\>标签

\<typeAliases\>标签的作用是为 Java 的 POJO 类起别名,如果不取别名,映射文件若要引用一个 POJO 实体类,必须输入全限定性类名,全限定性类名比较长,用了别名之后引用起来就简单多了。

例如，实体类 Student 的全限定性类名是 com.lifeng.entity.Student，未起别名之前，映射文件的 select 语句块若要引用 POJO 类 Student 必须使用全限定性类名，示例如下：

```xml
<select id="findAllStudents" resultType="com.lifeng.entity.Student">
        ELECT
        SELECT
            id as sid,
            studentname as sname,
            gender as sex,
            age
        FROM STUDENT
</select>
```

这里输入全路径

在主配置文件 mybatis-config.xml 中定义别名如下：

```xml
<typeAliases>
        <typeAlias alias="Student" type="com.lifeng.entity.Student"/>
</typeAliases>
```

上面代码的意思是为全限定类名 com.lifeng.entity.Student 创建别名 Student，定义了别名之后，映射文件中只须使用别名 Student 就能引用全限定类 com.lifeng.entity.Student。这样上例代码可简化如下：

```xml
<select id="findAllStudents" resultType="Student">
        SELECT
            id as sid,
            studentname as sname,
            gender as sex,
            age
        FROM STUDENT
</select>
```

用别名，更简洁

如果有多个类要使用别名，一种方法是逐个配置别名，多个 <typeAlias> 示例如下：

```xml
<typeAliases>
        <typeAlias alias="Student" type="com.lifeng.entity.Student"/>
        <typeAlias alias="Classes" type="com.lifeng.entity.Classes"/>
        <typeAlias alias="Course" type="com.lifeng.entity.Course"/>
        <typeAlias alias="Employee" type="com.lifeng.entity.Employee"/>
</typeAliases>
```

这样固然可以实现目的，但代码比较冗长，还有一个办法是通过自动扫描包的形式自定义别名，代码示例如下：

```xml
<typeAliases>
        <typeAlias name="com.lifeng.entity"/>
</typeAliases>
```

这样配置后，MyBatis 会自动扫描 <typeAlias> 节点的 name 属性指定的包 com.lifeng.entity，并自动将该包下的所有实体类以首字母为小写的类名作为别名，例如它会自动给 com.lifeng.entity.Student 分配一个别名 student，其他类推。

项目案例：使用别名简化了 MyBatis 映射文件。（项目源码参见本书配套资源：第 1 章/使用别名简化 MyBatis 映射文件/mybatis14）

实现步骤：
方法一：
（1）复制上一个项目 mybatis13 为 mybatis14，修改主配置文件，添加别名如下：

```xml
<?xml version="1.0" encoding="UTF-8"?>
```

```xml
<!DOCTYPE configuration PUBLIC "-//mybatis.org//DTD Config 3.0//EN"
"http://mybatis.org/dtd/mybatis-3-config.dtd">
 <configuration>
    <properties resource="jdbc.properties"/>
    <!-- 配置别名 第一种方法 直接指定别名-->
    <typeAliases>
        <typeAlias alias="Student" type="com.lifeng.entity.Student"/>
    </typeAliases>
    <environments default="development">
        <environment id="development">
            <transactionManager type="JDBC"/>
            <dataSource type="POOLED">
                <property name="driver" value="${jdbc.driver}"/>
                <property name="url" value="${jdbc.url}"/>
                <property name="username" value="${jdbc.username}"/>
                <property name="password" value="${jdbc.password}"/>
            </dataSource>
        </environment>
    </environments>
    <mappers>
        <mapper resource="com/lifeng/dao/StudentMapper.xml"/>
    </mappers>
</configuration>
```

（添加别名）

（2）修改映射文件 StudentMapper.xml，使用别名，代码如下：

```xml
<?xml version="1.0" encoding="utf-8" ?>
<!DOCTYPE mapper PUBLIC "-//mybatis.org//DTD Mapper 3.0//EN"
"http://mybatis.org/dtd/mybatis-3-mapper.dtd">
<mapper namespace="com.lifeng.dao.IStudentDao">
    <!-- 使用别名,理论上第一种方法 Student 用大写,第二种方法用小写，但测试表明大小写都能用 -->
    <select id="findAllStudents" resultType="Student">
        SELECT
            id as sid,
            studentname as sname,
            gender as sex,
            age
        FROM STUDENT
    </select>
</mapper>
```

（已简化，原来的是 com.lifeng.dao.Student）

（3）运行测试，结果不变。

方法二：

（1）修改主配置文件，用扫描包的方式配置别名，代码如下：

```xml
<?xml version="1.0" encoding="UTF-8"?>
<!DOCTYPE configuration PUBLIC "-//mybatis.org//DTD Config 3.0//EN"
"http://mybatis.org/dtd/mybatis-3-config.dtd">
<configuration>
    <!-- 注册属性文件 -->
    <properties resource="jdbc.properties"/>
    <typeAliases>
        <package name="com.lifeng.entity"/>
    </typeAliases>
    <environments default="development">
        <environment id="development">
            <transactionManager type="JDBC"/>
```

（用扫描包的方式批量将此包下的所有 POJO 类都自动设置别名，默认别名是类名首字母小写）

```xml
            <dataSource type="POOLED">
                <property name="driver" value="${jdbc.driver}"/>
                <property name="url" value="${jdbc.url}"/>
                <property name="username" value="${jdbc.username}"/>
                <property name="password" value="${jdbc.password}"/>
            </dataSource>
        </environment>
    </environments>
    <mappers>
        <mapper resource="com/lifeng/dao/StudentMapper.xml"/>
    </mappers>
</configuration>
```

（2）修改映射文件 StudentMapper.xml，使用别名，代码如下：

```xml
<?xml version="1.0" encoding="utf-8" ?>
<!DOCTYPE mapper PUBLIC "-//mybatis.org//DTD Mapper 3.0//EN"
"http://mybatis.org/dtd/mybatis-3-mapper.dtd">
<mapper namespace="com.lifeng.dao.IStudentDao">
    <select id="findAllStudents" resultType="student">
        SELECT
            id as sid,
            studentname as sname,
            gender as sex,
            age
        FROM STUDENT
    </select>
</mapper>
```

（已简化，原来是 com.lifeng.dao.Student）

（3）运行测试，结果不变。

1.7.3 <typeHandlers>标签

<typeHandlers>标签的作用是将传入参数的 javaType（Java 类型）转换为 jdbcType（JDBC 类型，对应数据库的数据类型），反之从数据库读出结果（即将 jdbcType 转换为 javaType）。

MyBatis 提供了大量默认的内置转换器，可以完成大部分常见的类型转换，但若默认的转换器无法满足要求时，可以自定义转换器。自定义转换器需实现 TypeHandler 接口或者继承 BaseTypeHandler 类。创建好后需要进行如下注册：

```xml
<typeHandlers>
    <typeHandler handler="com.lifeng.utils.MyTypeHandler"/>
</typeHandlers>
```

其中 handler 属性指定的类 com.lifeng.utils.MyTypeHandler 为自定义的转换器（类），然后<ResultMap>标签中就可以引用这个转换器了。

【注意】上面这段配置代码块必须放在 environments 前面和 typeAliases 后面。

自定义转换器实际使用起来比较复杂，还涉及后面的知识点，本书配套资源提供了一个完整项目案例 mybatis15 供参考，请读者至少学完第 2 章再回头来学习这个项目案例。

项目案例：复制项目 mybatis14 为 mybatis15，在数据库 student 表中添加一个列 address，类型为 VARCHAR，用于存储学生的地址信息，存储地址暂定用"城市,街道"这样的格式。在包 com.lifeng.entity 下新建一个类 Address，用于封装学生的地址信息，该类有 city 和 street 两个属性，重写 toString()方法，使城市 city 和街道 street 以","连接，创建带参数的构造方法，输入参数为"城市，街道"格式的字符串，在构造方法中进行拆分，分别给 city 和 street 属性赋值。实体类 Student 中添加一个 Address 类型的属性 address。实现学生信息的查找与添加。

问题分析：数据库中地址列是 VARCHAR 字符类型，Student 类中地址属性是 Address 对象类型，明显类型不一致，按之前的步骤，是无法正确查询并显示出数据来的，添加数据也一样。所以这里要用到自定义类型转换器，使数据库中的 address 列的 VARCHAR 类型能正确地转换映射到 Student 类的 address 属性的 Address 对象类型。（完整代码参见本书配套资源：第 1 章/typeHandler 自定义转换器/mybatis15）

关键步骤如下所示：

```java
package com.lifeng.utils;
import java.sql.CallableStatement;
import java.sql.PreparedStatement;
import java.sql.ResultSet;
import java.sql.SQLException;
import org.apache.ibatis.type.BaseTypeHandler;
import org.apache.ibatis.type.JdbcType;
import com.lifeng.entity.Address;
public class MyTypeHandler extends BaseTypeHandler<Address> {
    @Override
    public Address getNullableResult(ResultSet resultSet, String columnName)
            throws SQLException {
        return new Address(resultSet.getString(columnName));
    }

    @Override
    public Address getNullableResult(ResultSet resultSet, int columnIndex)
            throws SQLException {
        return new Address(resultSet.getString(columnIndex));
    }

    @Override
    public Address getNullableResult(CallableStatement callableStatement, int columnIndex)
            throws SQLException {
        return new Address(callableStatement.getString(columnIndex));
    }

    @Override
    public void setNonNullParameter(PreparedStatement preparedStatement, int index,
        Address address, JdbcType jdbcType) throws SQLException {
        preparedStatement.setString(index, address.toString());
    }
}
```

1.7.4 <environments>标签

<environments>标签是环境配置，主要用于数据源的配置。

1. <transactionManager>子标签

配置事务管理器可以配置 JDBC 和 MANAGED 两种类型。

① JDBC：使用 JDBC 的事务管理，通过数据源得到的连接来提交或回滚事务。

② MANAGED：使用容器来管理事务。

通常使用 JDBC 事务管理类型，配置方法如下：

```xml
<environments default="development">
    <environment id="development">
        <transactionManager type="JDBC"/>     ← 使用 JDBC 事务管理
        <dataSource type="POOLED">
            <property name="driver" value="${jdbc.driver}"/>
            <property name="url" value="${jdbc.url}"/>
            <property name="username" value="${jdbc.username}"/>
            <property name="password" value="${jdbc.password}"/>
        </dataSource>
    </environment>
</environments>
```

【注意】将来使用 Spring 整合 MyBatis 后，事务交给 Spring 管理，这部分也不用配置。

2. \<dataSource\>子标签的类型

\<dataSource\>标签用来配置数据源，即数据库的连接，它有 UNPOOLED、POOLED 和 JNDI 3 种类型。

（1）UNPOOLED

无连接池，每次请求都重新打开和关闭连接，即每一次都是新的连接，大型应用连接会很频繁，浪费资源，降低效率，一般只用于小型应用。

（2）POOLED

连接池，效率较高，响应速度快，通常都使用这种方式。

（3）JNDI

JNDI 数据源，常用于 EJB 等容器。

通常配置数据源类型为连接池，代码如下所示：

```xml
<environments default="development">
    <environment id="development">
        <transactionManager type="JDBC"/>
        <dataSource type="POOLED">     ← 配置数据源类型为连接池
            <property name="driver" value="${jdbc.driver}"/>
            <property name="url" value="${jdbc.url}"/>
            <property name="username" value="${jdbc.username}"/>
            <property name="password" value="${jdbc.password}"/>
        </dataSource>
    </environment>
</environments>
```

1.7.5 \<mappers\>标签

\<mappers\>标签用于配置映射文件，指定映射文件的位置有以下 4 种方法。

（1）使用 resource 属性引入类路径。示例如下：

```xml
<mappers>
    <mapper resource="com/lifeng/dao/StudentMapper.xml"/>
</mappers>
```

若配置文件有多个，可以按以下方法配置：

```xml
<mappers>
    <mapper resource="com/lifeng/dao/StudentMapper1.xml"/>
    <mapper resource="com/lifeng/dao/StudentMapper2.xml"/>
```

```
        <mapper resource="com/lifeng/dao/StudentMapper3.xml"/>
</mappers>
```
本书的案例就采用这种方法。

（2）使用 url 属性引入本地文件。示例如下：
```
<mappers>
        <mapper url="file:///E:/com/lifeng/dao/StudentMapper.xml"/>
</mappers>
```
url 也可以是网络地址，这样映射文件可以放在任何位置，但一般很少这样。

（3）使用 class 属性引入接口类。示例如下：
```
<mappers>
        <mapper class="com.lifeng.dao.StudentMapper"/>
</mappers>
```
这样配置比较简便，但要满足以下 3 个条件。

- 接口名称与映射文件名称一致。
- 接口与映射文件必须在同一个包中。
- 映射文件中<mapper>标签的 namespace 命名空间的值为接口的全限定类名。

本例中，接口名称必须为 StudentMapper，与映射文件放在同一个包 com.lifeng.dao 下，且映射文件<mapper>标签必须按以下方法配置：
```
<mapper namespace="com.lifeng.dao.StudentMapper">
```

（4）使用包名引入。

当映射文件较多时，可以不用一个个配置，而是使用如下形式，其中 package 的 name 属性指定映射文件所在的包，该包下的所有映射文件都会被扫描到，具体如下：
```
<mappers>
        <package name="com.lifeng.dao "/>
</mappers>
```
要使用这种配置，需满足以下 4 个条件。

- DAO 的实现类采用 mapper 动态代理实现。
- 映射文件与接口的名称相同。
- 映射文件与接口在同一个包中。
- 映射文件中<mapper>标签的 namespace 命名空间的值为接口的全限定类名。

1.8 连接其他数据库

1. 连接 Oracle 数据库

MyBatis 连接 Oracle 与 MySQL 的差别不大，只是驱动不同，连接字符串 URL 稍有不同。关键步骤如下。

（1）下载连接 Java 与 Oracle 的专用 JAR 包并导入项目。

（2）把 jdbc.properties 文件简单修改为如下格式：
```
jdbc.driver = oracle.jdbc.driver.OracleDriver
jdbc.url=jdbc:oracle:thin:@localhost:1521:orcl
jdbc.username=scott
jdbc.password=orcl
```

第 2 章有 MyBatis 连接 Oracle 的详细案例。

2. 连接 SQL Server 数据库

（1）下载连接 Java 与 SQL Server 的专用 JAR 包并导入项目。

（2）把 jdbc.properties 文件简单修改为如下格式：

```
jdbc.driver = com.microsoft.sqlserver.jdbc.SQLServerDrvier
jdbc.url= jdbc:sqlserver://127.0.0.1:1434;Database=Users
jdbc.username=admin
jdbc.password=123
```

上机练习

题目：在 MySQL 中创建表 1.1 所示的商品数据表 goods，在 Eclipse 中创建一个 MyBatis 项目，查询出全部商品。

表 1.1　商品数据表 goods

商品编号（id）	商品名称（goodsname）	商品单价（price）	商品数量（quantity）
1	电视机	5000	100
2	电冰箱	4000	200
3	空调	3000	300
4	洗衣机	3500	400

解题思路：模仿第一个 MyBatis 项目的实现步骤以及后续的各种优化（建议模仿项目 mybatis14）。自行创建数据库。

思考题

1. 简述 MyBatis 的工作流程。
2. MyBatis 的开发环境怎么搭建？
3. MyBatis 怎么配置别名？
4. 如何配置映射文件？

第 2 章　单表的增删改查

本章目标

- ✧ 理解结果映射 resultMap
- ✧ 掌握添加记录的方法
- ✧ 掌握获取自增长的 id 的方法
- ✧ 掌握修改记录的方法
- ✧ 掌握删除记录的方法
- ✧ 理解动态代理
- ✧ 掌握动态查询的方法
- ✧ 掌握 getMapper 面向接口编程的方法

第 1 章介绍了 MyBatis 可以方便地对数据库进行操作，而数据库表可能是一个相对独立的表（称为单表），也可能是两个或多个有密切联系的表（称为多表）。本章将介绍单表的增加、删除、修改、查询（增删改查）操作，下一章将介绍多表关联查询。

首先介绍 SqlSession 的若干方法，这些方法被用来执行定义在 mapper 映射文件中的 SELECT、INSERT、UPDATE 和 DELETE 语句。以下是一些常用的方法。

- <T> T selectOne(String statement, Object parameter)，查询单条记录。
- <E> List<E> selectList(String statement, Object parameter)，查询记录集合。
- <K,V> Map<K,V> selectMap(String statement, Object parameter, String mapKey)，查询记录封装成 Map 集合。
- int insert(String statement, Object parameter)，插入一条记录。
- int update(String statement, Object parameter)，修改一条记录。
- int delete(String statement, Object parameter)，删除一条记录。

上述方法中的第一个参数 statement 通常使用 namespace + id 的字符串，namespace 定位到唯一的 mapper 映射文件，id 定位到这个 mapper 映射文件指定的 SQL 语句。

第二个参数可以是原生类型（自动装箱或包装类）、JavaBean、POJO 或 Map，用于限定查询条件等。

其中 selectOne 和 selectList 的不同之处是 selectOne 必须返回一个对象，如果有多个对象，或者没有返回（或返回了 null），那么就会抛出异常。如果不确定返回了多少对象，就使用 selectList。

以上方法可以重载，第二个参数不是必要的，重载方法如下。

- <T> T selectOne(String statement)。
- <E> List<E> selectList(String statement)。
- <K,V> Map<K,V> selectMap(String statement, String mapKey)。
- int insert(String statement)。
- int update(String statement)。
- int delete(String statement)。

认识了这些方法后，下面来介绍实际操作，先将上面的项目 mybatis14 复制出另一个项目 mybatis21，在此基础上实现全部增删改查操作。

2.1 结果映射 resultMap

所谓结果映射就是让数据表的字段名称（列名）与 Java 实体类的属性名称一一进行关联匹配的一种机制，以便 MyBatis 查完数据库后能将关系型查询结果正确地封装为 Java 对象。如果数据库的字段名称跟相应的类的属性名称不一样，不处理的话，MyBatis 运行后相应的字段处的查询结果就为空，例如，在项目 mybatis21 的映射文件 StudentMapper.xml 中，如果 SQL 查询语句没有使用别名，即映射文件<select>节点中的 SQL 查询语句修改如下：

```
<select id="findAllStudents" resultType="com.lifeng.entity.Student">
    SELECT
        id,
        studentname,
```

```
            gender,
            age
        FROM STUDENT
</select>
```

运行后，结果如下：

```
DEBUG [main] - ==>  Preparing: SELECT id, studentname, gender, age FROM STUDENT
DEBUG [main] - ==> Parameters:
DEBUG [main] - <==      Total: 4
学生编号:null 学生姓名:null 学生性别 null 学生年龄:18
学生编号:null 学生姓名:null 学生性别 null 学生年龄:20
学生编号:null 学生姓名:null 学生性别 null 学生年龄:19
学生编号:null 学生姓名:null 学生性别 null 学生年龄:17
```

可以发现除了年龄，其他全变成了 null。

这是数据库表 student 的字段与对象 Student 的属性名称不一致造成的。数据库查出来的数据无法封装到对应的对象属性中去，除了 age 名称相同，其他三个都不同，所以只有 age 的数据，没有其他三个的数据。解决办法有两个，其中一个就是在 SQL 语句中使用别名，让查询结果字段与对象属性一致，项目 mybatis11 用的正是这个办法；还有一个解决办法就是利用结果映射。下面通过一个项目案例学习结果映射的有关技术。

项目案例：在 mybatis21 中，利用结果映射查询全部学生。（项目源码参见配套资源：第 2 章/结果映射/mybatis21）

实现步骤：

（1）修改映射文件 StdentMapper.xml 的查询语句，删除别名，代码如下：

```
<select id="findAllStudents" resultType="com.lifeng.entity.Student">
    SELECT
        id,studentname,gender,age
    FROM STUDENT
</select>
```

可以发现这些 SQL 语句没有别名，如果就这样下去，结果也会像上一个例子一样，有些字段查出来为空。而通过下面步骤的结果映射可以解决这个问题。

（2）配置结果映射 resultMap。

在映射文件 StdentMapper.xml 中新增结果映射的相关代码，代码如下：

```
<?xml version="1.0" encoding="utf-8" ?>
<!DOCTYPE mapper PUBLIC "-//mybatis.org//DTD Mapper 3.0//EN"
"http://mybatis.org/dtd/mybatis-3-mapper.dtd">
<mapper namespace="com.lifeng.dao.IStudentDao">
    <resultMap id="studentResultMap" type="com.lifeng.entity.Student">
        <id property="sid" column="id" />
        <result property="sname" column="studentname" />
        <result property="sex" column="gender" />
        <result property="age" column="age" />
    </resultMap>
    <select id="findAllStudents" resultType="Student">
        SELECT
            id,
            studentname,
            gender,
            age
```

（此处配置结果映射，id 为 studentResultMap）

（接下来要修改此处）

```
        FROM STUDENT
    </select>
</mapper>
```

这里 resultMap 的意思是把 SQL 查询结果映射为 Student 对象,其中数据库字段 id 映射为 Student 对象的 sid 属性,数据库字段 studentname 映射为 Student 对象的 sname 属性,数据库字段 gender 映射为 Student 对象的 sex 属性,数据库字段 age 映射为 Student 对象的 age 属性。

(3)修改映射文件中的查询语句。

将 resultType="Student"改为 resultMap="studentResultMap",其中,studentResultMap 为上一步配置的结果映射的 id 号,代码如下:

```
<select id="findAllStudents"  resultMap="studentResultMap">
    SELECT
        id,
        studentname,
        gender,
        age
    FROM STUDENT
</select>
```

原来是 resultType="Student"

(4)运行原有的测试类,代码如下:

```
DEBUG [main] - ==>  Preparing: SELECT id, studentname, gender, age FROM STUDENT
DEBUG [main] - ==> Parameters:
DEBUG [main] - <==      Total: 5
学生编号:1 学生姓名:张飞 学生性别:男 学生年龄:18
学生编号:2 学生姓名:李白 学生性别:男 学生年龄:20
学生编号:3 学生姓名:张无忌 学生性别:男 学生年龄:19
学生编号:4 学生姓名:赵敏 学生性别:女 学生年龄:17
```

可以正常输出了,没有出现 null 的情况,证明了结果映射成功。当然,如果映射文件中的查询语句修改为下面这种形式,也一样可以正常输出:

```
<select id=" findAllStudents "  resultMap="studentResultMap">
    SELECT *      FROM STUDENT
</select>
```

2.2 使用 selectOne 方法查询单条记录

项目案例:根据 id 号查询学生。(项目源码参见配套资源:第 2 章/增删改查/mybatis22)

实现步骤:

(1)将项目 mybatis21 复制成 mybatis22,在接口类 IStudentDao 中新增一个方法:
```
public Student findStudentById(int id);
```

(2)在映射文件 StudentMapper.xml 中新增一段 select 语句:
```
<select id="findStudentById"  parameterType="int"  resultMap="studentResultMap">
        SELECT * FROM STUDENT where id=#{id}
</select>
```

其中,parameterType="int"表示参数的类型为整型,也可省略不用,MyBatis 会自动识别传进来的参数类型。此外仍然使用结果映射 resultMap。

<select>标签中的 id 属性用于唯一标识该 SQL 语句,要与接口中方法名称一致。

SQL 语句中的#{ }：对指定参数类型属性值的引用。

（3）在接口 IStudentDao 的实现类 StudentDaoImpl 中新增方法，代码如下：

```
@Override
public Student findStudentById(int id) {
    Student student = new Student();
    try {
        session = MyBatisUtil.getSession();
        student = session.selectOne("com.lifeng.dao.IStudentDao.findStudentById", id);
    } catch (Exception e) {
        e.printStackTrace();
    }
    return student;
}
```

其中，跟上一个项目不同的地方是用 session.selectOne 取代了之前的 session.selectList 方法，这是 SqlSession 类的另一个方法，用于查询单条记录。还要注意这里的参数传递，形参 id 在方法 session.selectOne 的第二个参数中被调用到了。

- selectList 方法：适用于查询结果是多条的情形。
- selectOne 方法：适用于查询结果是单条的情形。

（4）在测试类 TestStudent1 中添加一个方法 findStudentById(int id)，代码如下：

```
public static void main(String[] args) {
    //findAllStudents();
    findStudentById(1);
}
public static void findStudentById(int id){
    IStudentDao studentDao=new StudentDaoImpl();
    Student student=studentDao.findStudentById(id);
    student.show();
}
```

结果如下：

```
DEBUG [main] - ==>  Preparing: SELECT id, studentname, gender, age FROM STUDENT where id=?
DEBUG [main] - ==> Parameters: 1(Integer)
DEBUG [main] - <==      Total: 1
学生编号:1 学生姓名:张飞 学生性别男 学生年龄:18
```

2.3 使用 insert 方法添加记录

2.3.1 主键非自增长

项目案例：在项目 mybatis22 中实现新增一个学生的功能，学号、姓名、性别、年龄均由程序给定。其中，学号在数据库中是主键（非自增长）。（项目源码参见配套资源：第 2 章/增删改查/mybatis22）

实现步骤：

（1）检查数据库 student 表，确认主键值非自增长，在接口类 IStudentDao 中新增一个方法。

```
public int insertStudent(Student student);
```

（2）在映射文件 StdentMapper.xml 中新增一段 insert 语句如下：

```xml
<!-- 添加一个学生 -->
<insert id="insertStudent" parameterType="Student">
    INSERT INTO student(id, studentname, gender, age)
    VALUES
      (#{sid}, #{sname}, #{sex}, #{age})
</insert>
```

parameterType="Student"也可不加，MyBatis 会自动判断传进来的参数类型。#{sid}是占位符，传进来的值是参数类型 Student 对应的同名属性值，即 Student 对象的 sid 属性的值，其他类推。

（3）在接口 IStudentDao 的实现类 StudentDaoImpl 中新增 insertStudent 方法如下：

```java
@Override
public int insertStudent(Student student) {
    SqlSession session = null;
    int count=0;
    try {
        session = MyBatisUtil.getSession();
        count = session.insert("com.lifeng.dao.IStudentDao.insertStudent",student);
    } catch (Exception e) {
        e.printStackTrace();
    }finally {
        session.close();
    }
    return count;
}
```

这里用到了 insert 方法，用于添加记录，返回受影响的行数。

（4）测试类中新增方法，代码如下：

```java
public static void main(String[] args) {
    //findAllStudents();
    //findStudentById(1);
    insertStudent();
}
//添加学生(给定主键值)
public static void insertStudent() {
    IStudentDao studentDao=new StudentDaoImpl();
    Student student=new Student();
    student.setSid("6");
    student.setSname("武松");
    student.setSex("男");
    student.setAge(23);
    int count=studentDao.insertStudent(student);
    if(count>0){
        System.out.println("添加成功!");
    }
}
```

测试结果如下：

```
DEBUG [main] - ==>  Preparing: INSERT INTO student(id, studentname, gender, age) VALUES (?, ?, ?, ?)
DEBUG [main] - ==> Parameters: 6(String), 武松(String), 男(String), 23(Integer)
DEBUG [main] - <==    Updates: 1
添加成功!
```

虽然当前提示添加成功，但查看数据库会发现数据中并没有这条新记录，原因是 update 后没有提交事务，所以还要提交事务。使用 session.commit()方法可实现事务提交。StudentDaoImpl 中的 insertStudent 方法修改如下：

```
@Override
public int insertStudent(Student student) {
    SqlSession session = null;
    int count=0;
    try {
        session = MyBatisUtil.getSession();
        count = session.insert("com.lifeng.dao.IStudentDao.insertStudent",student);
        session.commit();                          ← 提交事务
    } catch (Exception e) {
        e.printStackTrace();
    }finally {
        session.close();
    }
    return count;
}
```

再次测试，此时数据库就有新的记录了。

2.3.2 主键值由数据库自增长

MySQL 数据库表的主键值在设置为自动增长的情况下，程序只需要添加除主键外的其他字段即可。但一般情况下程序中添加一个 Student 对象后无法马上获得该对象的 id 属性值，需要一定的配置才行，具体步骤见下例。

项目案例：在项目 mybatis22 基础上，实现数据库 student 表主键自增长情况下的添加记录。（项目源码参见配套资源：第 2 章/增删改查/mybatis22）

实现步骤：

（1）修改 MySQL 数据库，设置 Student 表的主键 sid 自动增长（见图 2.1）。

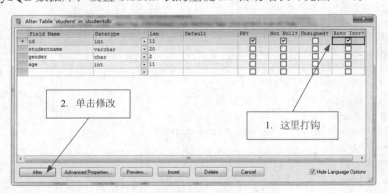

图 2.1 修改数据库表主键自增长

【注意】这里用的 MySQL 客户端是 SQLyog。

（2）在 StudentMapper.xml 中新增一条 insert 语句：

```xml
<insert id="insertStudentAutoIncrement" parameterType="Student" >
    INSERT INTO student(studentname, gender, age)
    VALUES
    (#{sname}, #{sex}, #{age})
</insert>
```

【注意】这里的 SQL 语句没有用到 sid 主键。

（3）在接口 IStudentDao 中添加方法 insertStudentAutoIncrement 如下：
```
public int insertStudentAutoIncrement(Student student);
```
（4）在接口的实现类 StudentDaoImpl 中重写方法 insertStudentAutoIncrement，代码如下：
```
@Override
public int insertStudentAutoIncrement(Student student) {
    SqlSession session = null;
    int count=0;
    try {
        session = MyBatisUtil.getSession();
        count = session.insert("com.lifeng.dao.IStudentDao.insertStudentAutoIncrement",student);
        session.commit();
    } catch (Exception e) {
        e.printStackTrace();
    }finally {
        session.close();
    }
    return count;
}
```
（5）第一次测试，代码如下：
```
public static void main(String[] args) {
    //findAllStudents();
    //findStudentById(1);
    //insertStudent();
    insertStudentAutoIncrement();
}
//添加学生(主键自增长)
public static void insertStudentAutoIncrement(){
    IStudentDao studentDao=new StudentDaoImpl();
    Student student=new Student();
    //第一次测试(不能获得主键值)
    student.setSname("林冲");
    student.setSex("男");
    student.setAge(22);
    int count=studentDao.insertStudentAutoIncrement(student);
    if(count>0){
        System.out.println("添加成功!");
        System.out.println("新添加的学生的编号是:"+student.getSid());
    }
}
```
结果如下：
```
DEBUG [main] - ==>  Preparing: INSERT INTO student(studentname, gender, age) VALUES (?, ?, ?)
DEBUG [main] - ==> Parameters: 林冲(String), 男(String), 22(Integer)
DEBUG [main] - <==    Updates: 1
添加成功!
新添加的学生的编号是:null
```
可见添加是成功的，但添加后学生编号为 null，即无法在添加完数据后立即获取主键的值，如何在数据添加成功后获取这个自增长的主键值呢？请看下一个步骤。

（6）修改映射文件，代码如下：
```xml
<insert id="insertStudentAutoIncrement" parameterType="Student"
keyProperty="sid" useGeneratedKeys="true">
        INSERT INTO student(studentname, gender, age)
        VALUES
           (#{sname}, #{sex}, #{age})
</insert>
```
关键代码是 keyProperty="sid" useGeneratedKeys="true"，其中，**keyProperty** 指的是实体类的某个属性接收主键返回的值，useGeneratedKeys 指的是使用自动生成的主键。

（7）第二次测试，代码如下：
```java
public static void main(String[] args) {
    //findAllStudents();
    //findStudentById(1);
    //insertStudent();
    insertStudentAutoIncrement();
}
//添加学生(主键自增长)
public static void insertStudentAutoIncrement(){
    IStudentDao studentDao=new StudentDaoImpl();
    Student student=new Student();
    //第二次测试(可以获得主键值)
    student.setSname("苏东坡");
    student.setSex("男");
    student.setAge(21);
    int count=studentDao.insertStudentAutoIncrement(student);
    if(count>0){
        System.out.println("添加成功!");
        System.out.println("新添加的学生的编号是:"+student.getSid());
    }
}
```

第二次测试结果如下：
```
DEBUG [main] - ==>  Preparing: INSERT INTO student(studentname, gender, age) VALUES (?, ?, ?)
DEBUG [main] - ==> Parameters: 苏东坡(String), 男(String), 21(Integer)
DEBUG [main] - <==    Updates: 1
添加成功!
新添加的学生的编号是:7
```

可以发现，已经获取到新添加的学生的主键值。

以上获取主键值的配置对 **MySQL** 和 **SQL Server** 数据库均有效，但对 **Oracle** 数据库无效，Oracle 数据库需要如下配置，假定 Oracle 数据库使用的序列名称为 seq_id，代码如下：
```xml
<insert id="insertStudentAutoIncrement" parameterType="Student"
   keyProperty="sid" useGeneratedKeys="true">
   <selectKey resultType="java.lang.Long" keyProperty="sid" order="before">
            select seq_id.nextval from dual
   </selectKey>
        INSERT INTO student(studentname, gender, age)
        VALUES
           (#{sname}, #{sex}, #{age})
</insert>
```

其中关键代码如下：
```
<selectKey resultType="java.lang.Long" keyProperty="sid" order="before">
        select seq_id.nextval from dual
</selectKey>
```
order="before"的意思是将序列返回数字转化为 Long 类型，该 select 语句比插入语句先执行，keyProperty="sid"的意思是将这个 Long 的值放入 parameterType="Student"的 sid 属性中。

2.4 使用 delete 方法删除记录

项目案例：在项目 mybatis22 中实现删除指定学号的学生的功能。（项目源码参见配套资源：第 2 章/增删改查/mybatis22）

实现步骤：

（1）在接口 IStudentDao 中添加一个方法 deleteStudentById(int id)如下：
```
public int deleteStudentById(int id);
```
（2）在映射文件 StudentMapper 中添加一条 delete 语句如下：
```
<delete id="deleteStudentById" parameterType="int">
     delete from student where id=#{id}
</delete>
```
（3）实现类 StudentDaoImpl，代码如下：
```
@Override
public int deleteStudentById(int sid) {
    SqlSession session = null;
    int count=0;
    try {
        session = MyBatisUtil.getSession();
        count = session.delete("com.lifeng.dao.IStudentDao.deleteStudentById",sid);
        session.commit();
    } catch (Exception e) {
        e.printStackTrace();
    }finally {
        session.close();
    }
    return count;
}
```
（4）在测试类中添加 deleteStudentById 方法如下：
```
public static void main(String[] args) {
    deleteStudentById();
}
public static void deleteStudentById(){
    IStudentDao studentDao=new StudentDaoImpl();
    Scanner input=new Scanner(System.in);
    System.out.println("请输入你要删除的学生的学号:");
    int sid=input.nextInt();
    Student student=studentDao.findStudentById(sid);
    if(student!=null){
        int count=studentDao.deleteStudentById(sid);
        if(count>0){
            System.out.println("删除成功!");
```

```
        }
    }else{
        System.out.println("该学生不存在!");
    }
}
```

结果如下：
请输入你要删除的学生的学号：
4
DEBUG [main] - ==> Preparing: delete from student where id=?
DEBUG [main] - ==> Parameters: 4(Integer)
DEBUG [main] - <==   Updates: 1
删除成功！

2.5　使用 update 方法修改记录

项目案例：在项目 mybatis22 中实现修改学生信息。（项目源码参见配套资源：第 2 章/增删改查/mybatis22）

实现步骤：

（1）在接口 IStudentDao 中添加一个方法 updateStudent(Student student)如下：
```
public int updateStudent(Student student);
```
（2）在映射文件 StudentMapper 中添加一条 update 语句如下：
```
<update id="updateStudent" parameterType="Student">
    UPDATE Student SET studentname=#{sname},gender=#{sex},age=#{age}
    WHERE id= #{sid}
</update>
```
（3）在实现类 StudentDaoImpl 中添加如下方法：
```
public int updateStudent(Student student){
    SqlSession session = null;
    int count=0;
    try {
        session = MyBatisUtil.getSession();
        count = session.update("com.lifeng.dao.IStudentDao.updateStudent",student);
        session.commit();
    } catch (Exception e) {
        e.printStackTrace();
    }finally {
        session.close();
    }
    return count;
}
```
（4）在测试类中添加 updateStudent 方法，代码如下：
```
public static void updateStudent(){
    IStudentDao studentDao=new StudentDaoImpl();
    Scanner input=new Scanner(System.in);
    System.out.println("请输入你要修改的学生的学号:");
    int sid=input.nextInt();
    Student student=studentDao.findStudentById(sid);
    if(student!=null){
        System.out.println("该学生原有信息如下");
```

```
            student.show();
            System.out.print("请输入学生的新的姓名:");
            String sname=input.next();
            System.out.print("请输入学生的新的性别:");
            String sex=input.next();
            System.out.print("请输入学生的新的年龄:");
            int age=input.nextInt();
            student=new Student();
                    student.setSname(sname);
                    student.setSex(sex);
                    student.setAge(age);
                    student.setSid(sid);
            studentDao.updateStudent(student);
    }else{
            System.out.println("该学生不存在");
    }
}
```

结果如下：

请输入你要修改的学生的学号：
4
DEBUG [main] - ==> Preparing: SELECT id, studentname, gender, age FROM STUDENT where id=?
DEBUG [main] - ==> Parameters: 4(Integer)
DEBUG [main] - <==   Total: 1
该学生原有信息如下
学生编号:4 学生姓名:赵敏 学生性别:女 学生年龄:17
请输入学生的新的姓名:赵敏
请输入学生的新的性别:女
请输入学生的新的年龄:18
DEBUG [main] - ==> Preparing: UPDATE student SET studentname=?,gender=?,age=? WHERE id= ?
DEBUG [main] - ==> Parameters: 赵敏(String), 女(String), 18(Integer), 4(String)
DEBUG [main] - <==   Updates: 1

2.6 模糊查询

项目案例：在项目 mybatis23 中实现查询姓张的学生。(项目源码参见配套资源：第 2 章/模糊查询、动态查询/mybatis23)

实现步骤：

(1) 将项目 mybatis22 复制为项目 mybatis23，在接口 IStudentDao 中添加一个 findStudentsByName(String name)方法如下：

```
public List<Student> findStudentsByName(String name);
```

(2) 在映射文件 StudentMapper 中添加一条 select 语句如下：

```
<select id="findStudentsByName" resultMap="studentResultMap" parameterType="String">
    SELECT * FROM STUDENT where studentname like '%' #{name} '%'
</select>
```

此外也可用如下语句实现一样的效果：

```
<select id="findStudentsByName" resultMap="studentResultMap" parameterType="String">
    SELECT * FROM STUDENT where studentname like concat('%', #{name}, '%')
</select>
```

即使用 concat 函数实现'%'、#{name}、'%'三者连接。

以上两种都是采用占位符的方式实现模糊查询,它们在控制台输出的 SQL 语句如下所示:
```
SELECT * FROM STUDENT where studentname like '%' ? '%'
```
可以发现 SQL 语句里面包含?符号,这正是我们熟悉的占位符。还有一种方式是拼接方式,见下面代码:
```
<select id="findStudentsByName" resultMap="studentResultMap" parameterType="String">
    SELECT * FROM STUDENT where studentname like '%${value}%'
</select>
```
要注意${}符号里面只能是 value,不能是其他单词。这种格式也能实现模糊查询,其控制台输出的 SQL 语句如下:
```
SELECT * FROM STUDENT where studentname like '%张%'
```
可见其 SQL 语句用的是硬编码拼接的方式,这种方式存在 SQL 注入漏洞,不推荐使用。

(3)在实现类中添加 findStudentsByName 方法,代码如下:
```
@Override
public List<Student> findStudentsByName(String name) {
    SqlSession session = null;
    List<Student> list = new ArrayList<Student>();
    try {
        session = MyBatisUtil.getSession();
        list=session.selectList("com.lifeng.dao.IStudentDao.findStudentsByName", name);
    } catch (Exception e) {
        e.printStackTrace();
    }
    return list;
}
```

(4)在测试类中添加方法 findStudentByName,代码如下:
```
public static void main(String[] args) {
    findStudentsByName();
}
public static void findStudentsByName(){
    Scanner input=new Scanner(System.in);
    System.out.print("请输入要查询的学生姓氏:");
    String name=input.next();
    IStudentDao studentDao=new StudentDaoImpl();
    List<Student> list=studentDao.findStudentsByName(name);
    for(int i=0;i<list.size();i++){
        list.get(i).show();
    }
}
```
结果如下:
```
请输入要查询的学生姓氏:张
DEBUG [main] - ==>  Preparing: SELECT * FROM STUDENT where studentname like '%' ? '%'
DEBUG [main] - ==> Parameters: 张(String)
DEBUG [main] - <==      Total: 2
学生编号:1 学生姓名:张飞 学生性别男 学生年龄:18
学生编号:3 学生姓名:张无忌 学生性别男 学生年龄:19
```
如果 SQL 语句采用拼接的方式,则结果如下:
```
请输入要查询的学生姓氏:张
```

```
DEBUG [main] - ==>  Preparing: SELECT * FROM STUDENT where studentname like '%张%'
DEBUG [main] - ==> Parameters:
DEBUG [main] - <==      Total: 2
学生编号:1 学生姓名:张飞 学生性别男 学生年龄:18
学生编号:3 学生姓名:张无忌 学生性别男 学生年龄:19
```

可以发现,尽管两者查询结果一样,但 SQL 语句不同,前者采用占位符,后者采用硬编码拼接的方式,后者有 SQL 注入漏洞。

2.7 动态查询

在大家熟悉的网购中,筛选商品是一个非常重要的功能。筛选条件有很多,每个人选的条件都不一样,这里就用到了动态查询。前台页面把某个用户的筛选条件传到后台服务器,后台服务器根据筛选条件动态生成 SQL 语句,再从数据库中查找到合适商品。

动态 SQL 多用于解决查询条件不确定的情况,在程序运行期间,根据用户提交的多种可能的查询条件进行查询,提交的查询条件不同,动态生成和执行的 SQL 语句也不同。动态 SQL 通过 MyBatis 提供的各种标签对条件做出判断以实现动态拼接 SQL 语句。

常用的动态 SQL 标签有<if>、<where>、<choose/>、<foreach>等。

2.7.1 <if>标签

对于该标签的执行,当 test 的值为 true 时,会将其包含的 SQL 片段拼接到其所在的 SQL 语句中。

项目案例:查询出满足用户查询条件的所有学生。(项目源码参见配套资源:第 2 章/模糊查询、动态查询/mybatis23)

问题分析:用户提交的查询条件可以包含一个姓名的模糊查询,同时还可以包含一个年龄的下限。当然,用户在提交表单时可能两个条件均做出了设定,也可能两个条件均不做设定,或者只做其中一项设定。需要解决的问题是查询条件不确定,且依赖于用户提交的内容。此时,就可使用动态 SQL 语句,根据用户提交的内容对将要执行的 SQL 进行拼接。

实现步骤:

(1)在接口 IStudentDao 中添加方法 searchStudentsIf(Student student)如下:
```
public List<Student> searchStudentsIf(Student student);
```

(2)在映射文件 StudentMapper.xml 中添加一条 select 语句,内容如下:
```
<select id="searchStudentsIf" resultMap="studentResultMap" parameterType="Student">
    SELECT * FROM STUDENT where 1=1
    <if test="sname!=null and sname!=''">
        and studentname like '%' #{sname} '%'
    </if>
    <if test="sex!=null and sex!=''">
        and gender=#{sex}
    </if>
    <if test="age>0">
        and age=#{age}
    </if>
</select>
```

if 子句的作用在于有条件地拼接 SQL 语句，如果条件成立，则拼接 SQL 语句，否则不拼接，所以最终的 SQL 语句是动态的。本例中，每个 if 子句的条件是否成立又取决于传过来的参数 Student 对象的属性是否为空。if 里面的判断条件使用的是 OGNL 表达式。

（3）在实现类中添加如下方法：

```java
@Override
public List<Student> searchStudentsIf(Student student) {
    SqlSession session = null;
    List<Student> list = new ArrayList<Student>();
    try {
        session = MyBatisUtil.getSession();
        list = session.selectList("com.lifeng.dao.IStudentDao.searchStudentsIf", student);
    } catch (Exception e) {
        e.printStackTrace();
    }
    return list;
}
```

（4）在测试类中添加 searchStudentIf 方法，代码如下：

```java
public static void main(String[] args) {
    searchStudentsIf();
}
public static void searchStudentsIf(){
    Scanner input=new Scanner(System.in);
    Student student=new Student();
    System.out.print("请输入要查询的学生姓名(也可回车跳过):");
    student.setSname(input.nextLine());
    System.out.print("请输入要查询的学生性别(也可回车跳过):");
    student.setSex(input.nextLine());
    System.out.print("请输入要查询的学生年龄(也可输入 0 跳过):");
    student.setAge(input.nextInt());
    IStudentDao studentDao=new StudentDaoImpl();
    List<Student> list=studentDao.searchStudentsIf(student);
    for(int i=0;i<list.size();i++){
        list.get(i).show();
    }
}
```

测试情形一：查询姓张的男生。结果如下：

请输入要查询的学生姓名(也可回车跳过):张
请输入要查询的学生性别(也可回车跳过):男
请输入要查询的学生年龄(也可输入 0 跳过):0
DEBUG [main] - ==> Preparing: SELECT * FROM STUDENT where 1=1 and studentname like '%' ? '%' and gender=?
DEBUG [main] - ==> Parameters: 张(String), 男(String)
DEBUG [main] - <==   Total: 2
学生编号:1 学生姓名:张飞 学生性别男 学生年龄:18
学生编号:3 学生姓名:张无忌 学生性别男 学生年龄:19

可看到 SQL 语句的 where 子句中使用了两个筛选条件。

测试情形二：查询 18 岁的女生。结果如下：

请输入要查询的学生姓名(也可回车跳过):

请输入要查询的学生性别(也可回车跳过):女
请输入要查询的学生年龄(也可输入 0 跳过):18
DEBUG [main] - ==> Preparing: SELECT * FROM STUDENT where 1=1 and gender=? and age=?
DEBUG [main] - ==> Parameters: 女(String), 18(Integer)
DEBUG [main] - <==   Total: 1
学生编号:4 学生姓名:赵敏 学生性别:女 学生年龄:18

可以发现 SQL 查询语句动态地改变了。

测试情形三：查询 18 岁的学生。结果如下：
请输入要查询的学生姓名(也可回车跳过):
请输入要查询的学生性别(也可回车跳过):
请输入要查询的学生年龄(也可输入 0 跳过):18
DEBUG [main] - ==> Preparing: SELECT * FROM STUDENT where 1=1 and age=?
DEBUG [main] - ==> Parameters: 18(Integer)
DEBUG [main] - <==   Total: 1
学生编号:1 学生姓名:张飞 学生性别男 学生年龄:18

2.7.2 \<where/>标签

<if/>标签中存在一个比较麻烦的地方，需要在 where 后手工添加"1=1"的子句。因为，若 where 后的所有<if/>条件均为 false，where 后没有 1=1 子句的话，则 SQL 中就会只剩下一个空的 where，SQL 出错。可以使用<where>标签实现同样的效果，而无须"1=1"这个看起来额外的东西。

项目案例：使用 where 标签查询出满足用户提交查询条件的所有学生。（项目源码参见配套资源：第 2 章/模糊查询、动态查询/mybatis23）

实现步骤：

（1）在接口 IStudentDao 中添加方法 searchStudentsWhere（Student student）如下。
```
public List<Student> searchStudentsWhere(Student student);
```

（2）在映射文件 StudentMapper.xml 中添加一条 select 语句，内容如下：
```xml
<select id="searchStudentsWhere" resultMap="studentResultMap" parameterType= "Student">
    SELECT * FROM STUDENT
    <where>
        <if test="sname!=null and sname!=''">
            and studentname like '%' #{sname} '%'
        </if>
        <if test="sex!=null and sex!=''">
            and gender=#{sex}
        </if>
        <if test="age>0">
            and age=#{age}
        </if>
    </where>
</select>
```

（3）在实现类中添加如下方法：
```java
//动态查询 where 标签
@Override
public List<Student> searchStudentsWhere(Student student) {
    SqlSession session = null;
    List<Student> list = new ArrayList<Student>();
    try {
```

```
            session = MyBatisUtil.getSession();
            list = session.selectList("com.lifeng.dao.IStudentDao.searchStudentsWhere",student);
        } catch (Exception e) {
            e.printStackTrace();
        }finally {
            session.close();
        }
        return list;
    }
```

（4）在测试类中添加如下方法：

```
public static void searchStudentsWhere(){
    Scanner input=new Scanner(System.in);
    Student student=new Student();
    System.out.print("请输入要查询的学生姓名(也可回车跳过):");
    student.setSname(input.nextLine());
    System.out.print("请输入要查询的学生性别(也可回车跳过):");
    student.setSex(input.nextLine());
    System.out.print("请输入要查询的学生年龄(也可输入0跳过):");
    student.setAge(input.nextInt());
    IStudentDao studentDao=new StudentDaoImpl();
    List<Student> list=studentDao.searchStudentsWhere(student);
    for(int i=0;i<list.size();i++){
        list.get(i).show();
    }
}
```

测试结果如下：

请输入要查询的学生姓名(也可回车跳过):张
请输入要查询的学生性别(也可回车跳过):男
请输入要查询的学生年龄(也可输入0跳过):0
DEBUG [main] - ==> Preparing: SELECT * FROM STUDENT WHERE studentname like '%' ? '%' and gender=?
DEBUG [main] - ==> Parameters: 张(String), 男(String)
DEBUG [main] - <==   Total: 2
学生编号:1 学生姓名:张飞 学生性别男 学生年龄:18
学生编号:3 学生姓名:张无忌 学生性别男 学生年龄:19

可见，同样实现了动态查询。

2.7.3　使用 Map 封装查询条件

上面两个案例中，多个查询条件都封装到一个 Student 对象的属性中去了，但在实际开发中，可能有些条件无法封装到同一个对象中去，这时就可以采用 Map 类型进行封装和传递参数。

项目案例：除上面的查询条件外，年龄可改为查询一定年龄范围内的学生。（项目源码参见配套资源：第 2 章/模糊查询动态、查询/mybatis23）

问题分析：这里有起始年龄、结束年龄，两个年龄无法封装到同一个 Student 对象中去。解决的办法就是将其封装到一个 Map 集合中。

实现步骤：

(1) 在接口 IStudentDao 中添加方法 searchStudentsMap（Map map）如下：
```
public List<Student> searchStudentsMap(Map map);
```

(2) 在映射文件 StudentMapper.xml 中添加一条 select 语句，内容如下：
```xml
<select id="searchStudentsMap" resultMap="studentResultMap" parameterType="java.util.Map">
    SELECT * FROM STUDENT
    <where>
        <if test="sname!=null and sname!=''">
            and studentname like '%' #{sname} '%'
        </if>
        <if test="sex!=null and sex!=''">
            and gender=#{sex}
        </if>
        <if test="ageStart>0 and ageEnd>0">
            and age between #{ageStart} and #{ageEnd}
        </if>
    </where>
</select>
```

上面代码的参数类型指定为 java.util.Map，SQL 语句中#{}里面的字符串对应 Map 集合的键（Key）名。

(3) 在实现类中添加如下方法：
```java
@Override
public List<Student> searchStudentsMap(Map map) {
    SqlSession session = null;
    List<Student> list = new ArrayList<Student>();
    try {
        session = MyBatisUtil.getSession();
        list = session.selectList("com.lifeng.dao.IStudentDao.searchStudentsMap", map);
    } catch (Exception e) {
        e.printStackTrace();
    }
    return list;
}
```

(4) 在测试类中添加如下方法：
```java
public static void main(String[] args) {
    searchStudentsMap();
}
public static void searchStudentsMap(){
    Scanner input=new Scanner(System.in);
    Map<String,Object> map=new HashMap<String,Object>();
    System.out.print("请输入要查询的学生姓名(也可回车跳过):");
    map.put("sname", input.nextLine());
    System.out.print("请输入要查询的学生性别(也可回车跳过):");
    map.put("sex", input.nextLine());
    System.out.print("请输入要查询的学生起始年龄(也可输入0跳过):");
    map.put("ageStart", input.nextInt());
    System.out.print("请输入要查询的学生结束年龄(也可输入0跳过):");
    map.put("ageEnd", input.nextInt());
    IStudentDao studentDao=new StudentDaoImpl();
```

```
       List<Student> list=studentDao.searchStudentsMap(map);
       for(int i=0;i<list.size();i++){
           list.get(i).show();
       }
}
```
上面代码把查询条件封装成了 Map 集合，这些查询条件是无法封装到一个 Student 对象中。

运行测试查询姓张的 18～19 岁的学生，测试结果如下：

请输入要查询的学生姓名(也可回车跳过):张
请输入要查询的学生性别(也可回车跳过):
请输入要查询的学生起始年龄(也可输入 0 跳过):18
请输入要查询的学生结束年龄(也可输入 0 跳过):19
DEBUG [main] - ==> Preparing: SELECT * FROM STUDENT WHERE studentname like '%' ? '%' and age between ? and ?
DEBUG [main] - ==> Parameters: 张(String), 18(Integer), 19(Integer)
DEBUG [main] - <==   Total: 2
学生编号:1 学生姓名:张飞 学生性别:男 学生年龄:18
学生编号:3 学生姓名:张无忌 学生性别:男 学生年龄:19

2.7.4 <choose/>标签

该标签里面包含<when/><otherwise/>两个子标签，可以包含多个<when/>与一个<otherwise/>。它们联合使用，可完成类似 Java 中的开关语句 switch…case 的功能。

项目案例：若姓名不为空，则按照姓名查询；若姓名为空，则按照年龄查询；若没有查询条件，则没有查询结果。（项目源码参见配套资源：第 2 章/模糊查询、动态查询/mybatis23）

实现步骤：

（1）在接口 IStudentDao 中添加方法 searchStudentsChoose(Student student)如下：
```
public List<Student> searchStudentsChoose(Student student);
```
（2）在映射文件 StudentMapper.xml 中添加一条 select 语句，内容如下：
```xml
<select id="searchStudentsChoose" resultMap="studentResultMap" parameterType= "Student">
    SELECT * FROM STUDENT
    <where>
        <choose>
            <when test="sname!=null and sname!=''">
                and studentname like '%' #{sname} '%'
            </when>
            <when test="age>0">
                and age=#{age}
            </when>
            <otherwise>
                and 1!=1
            </otherwise>
        </choose>
    </where>
</select>
```

<choose/>标签会从第一个<when/>开始，逐个向后进行条件判断。若出现<when/>中 test 属性值为 true 的情况，就直接结束<choose/>标签，不再向后进行判断查找。若所有<when/>的 test 判断结果均为 false，则最后会执行<otherwise/>标签。这种功能类似于 switch，实现了分支选择。

（3）在实现类中添加如下方法：
```java
@Override
public List<Student> searchStudentsChoose(Student student) {
    SqlSession session = null;
    List<Student> list = new ArrayList<Student>();
    try {
        session = MyBatisUtil.getSession();
        list = session.selectList("com.lifeng.dao.IStudentDao.searchStudentsChoose", student);
    } catch (Exception e) {
        e.printStackTrace();
    }finally {
        session.close();
    }
    return list;
}
```

（4）在测试类中添加如下方法：
```java
public static void main(String[] args) {
    searchStudentsChoose();
}
public static void searchStudentsChoose(){
    Scanner input=new Scanner(System.in);
    Student student=new Student();
    System.out.print("请输入要查询的学生姓名(也可回车跳过):");
    student.setSname(input.nextLine());
    System.out.print("请输入要查询的学生年龄(也可输入0跳过):");
    student.setAge(input.nextInt());
    IStudentDao studentDao=new StudentDaoImpl();
    List<Student> list=studentDao.searchStudentsChoose(student);
    for(int i=0;i<list.size();i++){
        list.get(i).show();
    }
}
```

测试情形一：姓名不为空，年龄为空（0）。

请输入要查询的学生姓名(也可回车跳过):张
请输入要查询的学生年龄(也可输入0跳过):0
DEBUG [main] - ==> Preparing: SELECT * FROM STUDENT WHERE studentname like '%' ? '%'
DEBUG [main] - ==> Parameters: 张(String)
DEBUG [main] - <==   Total: 2
学生编号:1 学生姓名:张飞 学生性别男 学生年龄:18
学生编号:3 学生姓名:张无忌 学生性别男 学生年龄:19

测试情形二：姓名为空，年龄不为0。

请输入要查询的学生姓名(也可回车跳过):
请输入要查询的学生年龄(也可输入0跳过):18
DEBUG [main] - ==> Preparing: SELECT * FROM STUDENT WHERE age=?
DEBUG [main] - ==> Parameters: 18(Integer)
DEBUG [main] - <==   Total: 1
学生编号:1 学生姓名:张飞 学生性别男 学生年龄:18

测试情形三：两者都为空（0）。

请输入要查询的学生姓名(也可回车跳过):

请输入要查询的学生年龄(也可输入0跳过):0
```
DEBUG [main] - ==>  Preparing: SELECT * FROM STUDENT WHERE 1!=1
DEBUG [main] - ==> Parameters:
DEBUG [main] - <==      Total: 0
```

2.7.5 使用<foreach/>标签遍历数组

<foreach/>标签用于实现对数组与集合类型的输入参数的遍历。其 collection 属性表示要遍历的集合类型，如果是数组，其值就用 array。open、close、separator 这些属性用于对遍历内容进行 SQL 拼接。具体用法请看下面的实例。

项目案例：查询出 id 为 1、3、4 的学生信息，利用数组作为参数存储 1、3、4。（项目源码参见配套资源：第 2 章/模糊查询、动态查询/mybatis23）

实现步骤：

（1）在接口 IStudentDao 中添加方法 searchStudentsEachArray（Object[] ids）如下：
```
public List<Student> searchStudentsEachArray(Object[] ids);
```
（2）在映射文件 StudentMapper.xml 中添加一条 select 语句，内容如下：
```xml
<select id="searchStudentsEachArray" resultMap="studentResultMap">
    SELECT * FROM STUDENT
    <if test="array!=null and array.length>0">
        where id in
        <foreach collection="array" open="(" close=")" item="myid" separator=",">
            #{myid}
        </foreach>
    </if>
</select>
```
其中，array 表示数组，匹配传进来的参数是数组类型，上述代码的意思是用 foreach 循环遍历数组，拼接出类似 in (1,3,4)格式的 SQL 语句。open="("表示以符号"("作为拼接字符串的开始，close=")"表示以符号")"作为拼接字符串的结束。separator=","表示分隔符号。item="myid"表示遍历的每一个元素可以自定义，但要跟<foreach></foreach>内部的占位符#{myid}一致。

（3）在实现类中添加如下方法：
```java
public List<Student> searchStudentsEachArray(Object[] ids){
    SqlSession session = null;
    List<Student> list = new ArrayList<Student>();
    try {
        session = MyBatisUtil.getSession();
        list = session.selectList("com.lifeng.dao.IStudentDao. searchStudentsEachArray", ids);
    } catch (Exception e) {
        e.printStackTrace();
    }finally {
        session.close();
    }
    return list;
}
```
（4）在测试类中添加如下方法：
```java
public static void main(String[] args) {
    searchStudentsEachArray();
}
```

```java
public static void searchStudentsEachArray(){
    Scanner input=new Scanner(System.in);
    System.out.print("请输入要查询的学生id(多个id之间用逗号隔开):");
    String id=input.next();
    String[] ids=id.split(",");
    IStudentDao studentDao=new StudentDaoImpl();
    List<Student> list=studentDao.searchStudentsEachArray(ids);
    for(int i=0;i<list.size();i++){
        list.get(i).show();
    }
}
```

测试结果如下：

请输入要查询的学生id(多个id之间用逗号隔开):1,3,4
DEBUG [main] - ==> Preparing: SELECT * FROM STUDENT where id in (? , ? , ?)
DEBUG [main] - ==> Parameters: 1(String), 3(String), 4(String)
DEBUG [main] - <==   Total: 3
学生编号:1 学生姓名:张飞 学生性别男 学生年龄:18
学生编号:3 学生姓名:张无忌 学生性别男 学生年龄:19
学生编号:4 学生姓名:赵敏 学生性别女 学生年龄:17

2.7.6 使用<foreach/>标签遍历泛型为基本类型的 List

项目案例：查询出 id 为 2 和 4 的学生信息。利用基本类型泛型集合作为参数存储 2 和 4。（项目源码参见配套资源：第 2 章/模糊查询、动态查询/mybatis23）

实现步骤：

（1）在接口 IStudentDao 中添加方法 searchStudentsEachList(List<Integer> list)如下：

```java
public List<Student> searchStudentsEachList(List<Integer> list);
```

（2）在映射文件 StudentMapper.xml 中添加一条 select 语句，内容如下：

```xml
<select id="searchStudentsEachList" resultMap="studentResultMap">
        SELECT * FROM STUDENT
        <if test="list!=null and list.size>0">
            where id in
            <foreach collection="list" open="(" close=")" item="myid" separator=",">
              #{myid}
            </foreach>
        </if>
</select>
```

其中，list 表示泛型，匹配传进来的参数是泛型类型，上述代码的意思是用 foreach 循环遍历泛型，拼接出类似 in (2,4)格式的 SQL 语句。

（3）在实现类中添加如下方法：

```java
public List<Student> searchStudentsEachList(List<Integer> idlist){
    SqlSession session = null;
    List<Student> list = new ArrayList<Student>();
    try {
        session = MyBatisUtil.getSession();
        list=session.selectList("com.lifeng.dao.IStudentDao.searchStudentsEachList",idlist);
    } catch (Exception e) {
        e.printStackTrace();
    }finally {
```

```
            session.close();
        }
        return list;
    }
```
（4）在测试类中添加如下方法：
```java
public static void main(String[] args) {
    searchStudentsEachList();
}
public static void searchStudentsEachList(){
    Scanner input=new Scanner(System.in);
    System.out.print("请输入要查询的学生id(多个id之间用逗号隔开):");
    String id=input.next();
    String[] ids=id.split(",");
    List<Integer> idlist=new ArrayList<Integer>();
    for(int i=0;i<ids.length;i++){
        idlist.add(Integer.parseInt(ids[i]));
    }
    IStudentDao studentDao=new StudentDaoImpl();
    List<Student> list=studentDao.searchStudentsEachList(idlist);
    for(int i=0;i<list.size();i++){
        list.get(i).show();
    }
}
```
测试结果如下：
```
请输入要查询的学生id(多个id之间用逗号隔开):2,4
DEBUG [main] - ==>  Preparing: SELECT * FROM STUDENT where id in ( ? , ? )
DEBUG [main] - ==> Parameters: 2(Integer), 4(Integer)
DEBUG [main] - <==      Total: 2
学生编号:2 学生姓名:李白 学生性别男 学生年龄:20
学生编号:4 学生姓名:赵敏 学生性别女 学生年龄:17
```

2.7.7　使用<foreach/>标签遍历泛型为自定义类型的 List

项目案例：查询出 id 为 2 和 4 的学生信息，利用 Student 类型泛型集合作为参数存储 2 和 4。（项目源码参见配套资源：第 2 章/模糊查询、动态查询/mybatis23）

实现步骤：

（1）在接口 IStudentDao 中添加方法 searchStudentsEachListStu(List<Student> list)如下：
```java
public List<Student> searchStudentsEachListStu(List<Student> list);
```
（2）在映射文件 StudentMapper.xml 中添加一条 select 语句，内容如下：
```xml
<select id="searchStudentsEachListStu" resultMap="studentResultMap">
    SELECT * FROM STUDENT
    <if test="list!=null and list.size>0">
        where id in
        <foreach collection="list" open="(" close=")" item="stu" separator=",">
         #{stu.sid}
        </foreach>
    </if>
</select>
```
（3）在实现类中添加如下方法：
```java
public List<Student> searchStudentsEachListStu(List<Student> stulist){
```

```
        SqlSession session = null;
        List<Student> list = new ArrayList<Student>();
        try {
            session = MyBatisUtil.getSession();
            list=session.selectList("com.lifeng.dao.IStudentDao.
searchStudentsEachListStu",stulist);
        } catch (Exception e) {
            e.printStackTrace();
        }finally {
            session.close();
        }
        return list;
    }
```

（4）在测试类中添加如下方法：

```
public static void main(String[] args) {
    searchStudentsEachListStu();
}
public static void searchStudentsEachListStu(){
    Student stu1=new Student();
    stu1.setSid("2");
    Student stu2=new Student();
    stu2.setSid("4");
    List<Student> stulist=new ArrayList<Student>();
    stulist.add(stu1);
    stulist.add(stu2);
    IStudentDao studentDao=new StudentDaoImpl();
    List<Student> list=studentDao.searchStudentsEachListStu(stulist);
    for(int i=0;i<list.size();i++){
        list.get(i).show();
    }
}
```

测试结果如下：

```
DEBUG [main] - ==>  Preparing: SELECT * FROM STUDENT where id in ( ? , ? )
DEBUG [main] - ==> Parameters: 2(String), 4(String)
DEBUG [main] - <==      Total: 2
学生编号:2 学生姓名:李白 学生性别男 学生年龄:20
学生编号:4 学生姓名:赵敏 学生性别女 学生年龄:17
```

2.7.8 <sql/>标签

<sql/>标签用于定义 SQL 片段供其他 SQL 标签复用。其他标签要使用该 SQL 片段，需要使用<include/>子标签。该<sql/>标签可以定义 SQL 语句中的任何部分，所以<include/>子标签可以放在动态 SQL 的任何位置，最终拼接出需要的 SQL 语句。

<sql/>标签的使用如下代码所示。（项目源码参见配套资源：第 2 章/模糊查询、动态查询/mybatis23）

```xml
<!-- 定义 SQL 片段 -->
<sql id="selectStu">
    SELECT * FROM STUDENT
</sql>                          ← SQL 片段

<select id="searchStudentsEachListStuSql" resultMap="studentResultMap">
```

```xml
            <!-- 使用SQL片段 -->
            <include refid="selectStu"/>          ← 调用SQL片段
            <if test="list!=null and list.size>0">
                where id in
                <foreach collection="list" open="(" close=")" item="stu" separator=",">
                    #{stu.sid}
                </foreach>
            </if>
</select>
```

2.8 分页查询基础

分页查询的原理在于两个参数，一个是页码 pageno，一个是页尺寸 pagesize，只要这两个参数确定，下面这个伪 SQL 语句就可查询出第 pageno 页的数据。

```
select * from student limit (pageno-1)*pagesize,pagesize
```

如果直接以 pageno 和 pagesize 作为参数传递，则 MySQL 数据库无法识别表达式 (pageno-1)*pagesize，所以通常把它综合为一个参数 startRow，让表达式(pageno-1)*pagesize 在传参前先计算好，赋值给参数 startRow。这样，最终使用的 SQL 语句如下所示：

```
select * from student limit startRow,pagesize
```

体现在映射文件中是下面的形式：

```
select * from student limit #{startRow},#{pagesize}
```

项目案例：分页查询出第 1 页、第 2 页的学生表数据，页面大小自定。（项目源码参见配套资源：第 2 章/模糊查询、动态查询/mybatis23）

实现步骤：

（1）在接口 IStudentDao 中添加方法 searchStudentsEachListStu（List<Student> list）如下：

```java
public List<Student> findAllStudentsByPage(Map map);
```

分页要用到两个参数，但由于 selectList 方法只能传递一个参数，所以这里采用了 Map 类型封装多个参数的办法。

（2）在映射文件 StudentMapper.xml 中添加一条 select 语句，内容如下：

```xml
<select id="findAllStudentsByPage" resultMap="studentResultMap" parameterType="java.util.Map">
    SELECT
        *
    FROM STUDENT limit #{startRow},#{pagesize}
</select>
```

其中，#{startRow}参数在实际传递的时候要先计算好(pageno-1)*pagesize 的值。

（3）在实现类中添加如下方法：

```java
public List<Student> findAllStudentsByPage(Map map){
    SqlSession session = null;
    List<Student> list = new ArrayList<Student>();
    try {
        session = MyBatisUtil.getSession();
        list = session.selectList("com.lifeng.dao.IStudentDao.findAllStudentsByPage", map);
    } catch (Exception e) {
        e.printStackTrace();
```

```
        }finally {
            session.close();
        }
        return list;
    }
```

（4）在测试类中添加如下方法：
```
    public static void main(String[] args) {
        findAllStudentsByPage();
    }
    public static void findAllStudentsByPage(){
        Scanner input=new Scanner(System.in);
        IStudentDao studentDao=new StudentDaoImpl();
        System.out.println("你想设置每页显示几条数据:");
        int pagesize=input.nextInt();
        System.out.println("你想查询第几页:");
        int pageno=input.nextInt();
        Map map=new HashMap();
        int startRow=(pageno-1)*pagesize;//先计算好起始行
        map.put("startRow", startRow);
        map.put("pagesize", pagesize);
        List<Student> sList=studentDao.findAllStudentsByPage(map);
        for(int i=0;i<sList.size();i++){
            sList.get(i).show();
        }
    }
```

通过 map.put("startRow ", startRow);这条语句封装参数"startRow"的实参的值，这样传的是计算好的(pageno-1)*pagesize 的值。

测试情形一：查询第 1 页，每页显示 2 条数据。
你想设置每页显示几条数据:
2
你想查询第几页:
1
DEBUG [main] - ==> Preparing: SELECT * FROM STUDENT limit ?,?
DEBUG [main] - ==> Parameters: 0(Integer), 2(Integer)
DEBUG [main] - <==   Total: 2
学生编号:1 学生姓名:张飞 学生性别男 学生年龄:18
学生编号:2 学生姓名:李白 学生性别男 学生年龄:20

测试情形二：查询第 2 页，每页显示 2 条数据。
你想设置每页显示几条数据:
2
你想查询第几页:
2
DEBUG [main] - ==> Preparing: SELECT * FROM STUDENT limit ?,?
DEBUG [main] - ==> Parameters: 2(Integer), 2(Integer)
DEBUG [main] - <==   Total: 2
学生编号:3 学生姓名:张无忌 学生性别男 学生年龄:19
学生编号:4 学生姓名:赵敏 学生性别女 学生年龄:17

测试情形三：查询第 1 页，每页显示 3 条数据。
你想设置每页显示几条数据:
3

你想查询第几页：
1
DEBUG [main] - ==> Preparing: SELECT * FROM STUDENT limit ?,?
DEBUG [main] - ==> Parameters: 0(Integer), 3(Integer)
DEBUG [main] - <==   Total: 3
学生编号:1 学生姓名:张飞 学生性别男 学生年龄:18
学生编号:2 学生姓名:李白 学生性别男 学生年龄:20
学生编号:3 学生姓名:张无忌 学生性别男 学生年龄:19

2.9　getMapper 面向接口编程

在前面例子中自定义 DAO 层接口实现类时发现一个问题：DAO 层的实现类其实并没有干什么实质性的工作，它仅仅就是通过 SqlSession 的相关 API 定位到映射文件 mapper 中相应 id 的 SQL 语句，真正对数据进行操作的工作其实是由框架通过映射文件 mapper 中的 SQL 完成的。

那么，MyBatis 框架可以抛开 DAO 的实现类，直接定位到映射文件 mapper 中的相应 SQL 语句，对数据库进行操作吗？答案是肯定的。可以通过 Mapper 的动态代理方式实现。下面结合案例来学习这种方式。

项目案例：无接口的实现类，通过动态代理，实现查找全部学生，以及查找 id="1"的学生。（项目源码参见配套资源：第 2 章/面向接口编程/mybatis24）

思路分析：上面项目已具备各种基本操作，只需按下面步骤简单修改即可。

实现步骤：

（1）复制项目 mybatis23 为 mybatis24，删除实现类 StudentDaoImpl。

（2）确保映射文件的 namespace 与接口类的全路径名称一致。接口类全路径名如下：
com.lifeng.dao.IStudentDao
映射文件命名空间 namespace 如下：
`<mapper namespace="com.lifeng.dao.IStudentDao">`

（3）新建测试类 TestStudent1。

Mapper 动态代理方式无须程序员实现 DAO 接口，DAO 实现对象是由 JDK 的 Proxy 动态代理自动生成的。使用时，只需调用 SqlSession 的 getMapper()方法，即可获取指定接口的实现类对象。该方法的参数为指定 DAO 接口类的 class 值，代码如下：

```java
public class TestStudent1 {
    public static void main(String[] args) {
        System.out.println("--------查询全部学生---------");
        findAllStudents();
        System.out.println("--------查询单个学生---------");
        findStudentById(1);
    }
    //查询全部学生,不带参
    public static void findAllStudents(){
        SqlSession session=MyBatisUtil.getSession();
        IStudentDao studentDao=session.getMapper(IStudentDao.class);
        List<Student> sList=studentDao.findAllStudents();
        for(int i=0;i<sList.size();i++){
            sList.get(i).show();
```

```
        }
    }
    //查询单个学生,带参
    public static void findStudentById(int id){
        SqlSession session=MyBatisUtil.getSession();
        IStudentDao studentDao=session.getMapper(IStudentDao.class);
        Student student=studentDao.findStudentById(id);
        student.show();
    }
}
```

从上面代码可以看到，通过 getMapper 方法即可获得接口的代理实现类，然后接口的各种方法都可直接调用。体现在 studentDao 中只要后面打个点，可以观察到所有接口的方法都能调用到。这个案例完全没用到 DAO 接口的实现类。

2.10 多参数查询

上面的案例都是只有一个参数的情况，或者虽然要传多个参数，但把这些参数都巧妙地封装成一个对象或一个 Map，表面上还是一个参数。有时想直接传递多个参数，不想封装，该如何实现呢？

项目案例：查询年龄为 18 岁的女学生。（项目源码参见配套资源：第 2 章/面向接口编程/mybatis24）
实现步骤：
（1）在接口 IStudentDao 中添加如下方法：
```
public List<Student> searchStudents(@Param("age")Integer age,@Param("sex")String gender)
```
这里用到了 @Param 注解。映射文件的 SQL 语句中的输入参数应与注解里面的参数一致。
（2）在 StudentMapper.xml 映射文件中添加如下 SQL 语句：
```
<!-- 多参数查询 -->
<select id="searchStudents" resultMap="studentResultMap">
    Select * from student where age=#{age} and gender=#{sex}
</select>
```
（3）在测试类 TestStudent1 类中实现上述接口，代码如下：
```
//查询 XX 岁的 X 性别的学生
    public static void findStudents(int age,String sex){
        SqlSession session=MyBatisUtil.getSession();
        IStudentDao studentDao=session.getMapper(IStudentDao.class);
        List<Student> sList=studentDao.searchStudents(age, sex);
        for(int i=0;i<sList.size();i++){
            sList.get(i).show();
        }
    }
```
主方法调用如下：
```
public static void main(String[] args) {
        System.out.println("--------查询18岁的女学生---------");
        findStudents(18,"女");
    }
```
查询结果如下：

```
--------查询18岁的女学生---------
DEBUG [main] - ==>  Preparing: Select * from student where age=? and gender=?
DEBUG [main] - ==> Parameters: 18(Integer), 女(String)
DEBUG [main] - <==      Total: 1
学生编号:23 学生姓名:西门吹水 学生性别:女 学生年龄:18
```

2.11 MyBatis 读写 Oracle 大对象数据类型

在项目中通常要把照片、视频、论文等资源保存到数据中，一般的类型保存不了，要用到 BLOB 或 CLOB 类型，Java 怎样实现这些类型数据的读写呢？下面用一个实例来学习这个技术。这里用到了 Oracle 数据库，MySQL 数据库也是适用的。

项目案例：读出 Oracle 数据库中一个学生的信息，包括其相片与简历，相片是 BLOB 类型的，简历是 CLOB 类型的，由于控制台无法显示图片，可将其用 IO 流存储到硬盘上。（项目源码参见配套资源：第 2 章/oracle 大数据读写/mybatis25）

思路分析：将数据库中的相片保存为二进制的 BLOB 类型，Java 中 POJO 类对应的属性为 byte[] 类型，这样就存在一个转换问题。个人简历在数据库中为 CLOB 类型，Java 中 POJO 类对应的属性为 String 类型，这个可以直接获取，无须转换。

实现步骤：

（1）Oracle 数据库的用户名为 scott，密码为 orcl，使用 PLSQL Developer 工具创建表和数据，具体如下：

```
create table STUDENT
(
  id       NUMBER not null,
  name     VARCHAR2(20),
  pic      BLOB,
  resume   CLOB
)
alter table STUDENT
  add constraint PK_ID primary key (ID)
```

数据请参考图 2.2 所示内容。

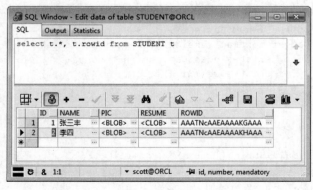

图 2.2 原始数据

其中每一次记录的 PIC 列旁边都有一个按钮，单击此按钮，进入图 2.3 所示界面，即可添加图片。

图 2.3　添加 BLOB 数据

单击每行记录的 RESUME 列中的 CLOB 按钮，即可添加大文本，如图 2.4 所示。

图 2.4　添加 CLOB 数据

（2）将项目 mybatis23 复制一份，命名为 mybatis25，修改 jdbc.properties 文件如下：

```
jdbc.driver=oracle.jdbc.driver.OracleDriver
jdbc.url=jdbc:oracle:thin:@localhost:1521:orcl
jdbc.username=scott
jdbc.password=orcl
```

可以发现，其连接的是 Oracle 数据库，用户名 scott，密码 orcl，相应地在 WebContent/WEB-INF/lib 下要添加 ojdbc6.jar 包并配置，可自行下载，或者在随书资源中查找相应项目。

（3）重新编写 Student 实体类，代码如下：

```
package com.lifeng.entity;
public class Student implements java.io.Serializable{
    private int id;
    private String name;
    private byte[] pic;
    private String resume;
    //省略setter,getter方法
}
```

重点是数据库中的是 BLOB 类型，对应实体类的属性用 byte[]类型，数据库中的 CLOB 类型，

49

对应实体类的属性用 String 类型。

（4）修改 IStudentDao 接口，仅保留查找与添加两个方法，其他方法删除，具体如下：

```
public interface IStudentDao {
    public Student findStudentById(int id);
    public int insertStudent(Student stu);
}
```

（5）修改 StudentMapper 接口，仅保留查找与添加两个 SQL 功能模块，其他 SQL 功能模块删除，此外<resultMap>节点是重点，留意看下面的代码：

```
<?xml version="1.0" encoding="utf-8" ?>
<!DOCTYPE mapper PUBLIC "-//mybatis.org//DTD Mapper 3.0//EN"
"http://mybatis.org/dtd/mybatis-3-mapper.dtd">
<mapper namespace="com.lifeng.dao.IStudentDao">
    <resultMap id="studentResultMap" type = "com.lifeng.entity.Student">
          <id property="id" column="id" />
           <result property="name" column="name" />
```

重点

```
          <result property="pic" column="pic" jdbcType="BLOB" javaType ="[B"/>
          <result   property="resume"   column="resume"   jdbcType="CLOB"   javaType   =
"java.lang.String"/>
    </resultMap>
    <!-- 下面的叫参数映射，适用于添加 insert 语句中一些特殊的对象类型与数据库类型之间的映射，本案例并
未用到，仅供参考 -->
    <!--  <parameterMap id="userPara" type = "com.lifeng.entity.Student">
        <parameter property="id" jdbcType="INTEGER" javaType ="java.lang.Integer"/>
        <parameter property="name" jdbcType="VARCHAR" javaType ="java.lang.String"/>
        <parameter property="pic" jdbcType="BLOB" javaType ="[B"/>
        <parameter property="resume" jdbcType="CLOB" javaType ="java.lang.String"/>
      </parameterMap>   -->
    <select id="findStudentById"  parameterType="int"  resultMap="studentResultMap">
          SELECT
               *
          FROM STUDENT where id=#{id}
    </select>
    <insert id="insertStudent" parameterType="Student" >
          INSERT INTO student
          VALUES
           (#{id}, #{name}, #{pic}, #{resume})
   </insert>
</mapper>
```

<result property="pic" column="pic" jdbcType="BLOB" javaType ="[B"/>这个配置比之前多了两个属性，resultMap 节点的作用除了之前熟知的结果映射功能，还实现了数据类型的转换。本例的难点在于数据库表 pic 字段的类型是 BLOB，而对应的 Student 对象的 pic 属性的类型是 byte[]，需要转换。<result>节点中 jdbcType="BLOB"表示数据库中的字段类型是 BLOB，javaType ="[B"表示映射对应的 Java 对象的属性的类型是 byte[]，整行语句实现了 BLOB 类型向 byte[]类型的转换映射功能。注意，byte[]类型有特殊的写法 javaType ="[B"，否则不能成功。同理 jdbcType="CLOB"表示数据库中的字段类型是 CLOB，javaType = "java.lang.String"表示映射对应的 Java 对象的属性的类型是 String，整行语句实现了 CLOB 类型向 String 类型的转换映射功能。

【注意】由于 MyBatis 内置了 BLOB 类型向 byte[]类型的转换器，以及 CLOB 类型向 String 类型的转换器，所以即使<result>节点不这样配置也可以，即下面这样简化配置也合法。

```xml
<resultMap id="studentResultMap" type = "com.lifeng.entity.Student">
    <id property="id" column="id" />
    <result property="name" column="name" />
    <result property="pic" column="pic" />
    <result property="resume" column="resume" />
</resultMap>
```

（6）接口的实现类只有以下两个方法，将原来多余的方法删除，代码如下：

```java
public class StudentDaoImpl implements IStudentDao{
    @Override
    public Student findStudentById(int id) {
        SqlSession session = null;
        Student student = new Student();
        try {
            session = MyBatisUtil.getSession();
            student = session.selectOne("com.lifeng.dao.IStudentDao.findStudentById", id);
        } catch (Exception e) {
            e.printStackTrace();
        }finally {
            session.close();
        }
        return student;
    }
    @Override
    public int insertStudent(Student student) {
        SqlSession session = null;
        int count=0;
        try {
            session = MyBatisUtil.getSession();
            count    =    session.insert("com.lifeng.dao.IStudentDao.insertStudent", student);
            session.commit();
        } catch (Exception e) {
            e.printStackTrace();
        }finally {
            session.close();
        }
        return count;
    }
}
```

（7）在测试类中添加如下方法：

```java
public class TestStudent1 {
    public static void main(String[] args) throws IOException, SQLException {
        findStudentById();
        //insertStudent();
    }
    public static void findStudentById() throws IOException, SQLException{
        IStudentDao studentDao=new StudentDaoImpl();
        //从数据库中读出数据封装到 student 对象中
        Student student=studentDao.findStudentById(1);
        //读取 pic 属性中的二进制数据
        byte[] pic=student.getPic();
        String resume=student.getResume();
        //创建输出流
```

```java
            OutputStream os=new FileOutputStream("c:\\mypic.jpg");
            //写出输出流,从而将数据库中的二进制数据转化为图片
            os.write(pic);
            System.out.println(resume);
        }
        public static void insertStudent() {
            IStudentDao studentDao=new StudentDaoImpl();
            Student student=new Student();
            student.setId(3);
            student.setName("西门吹雪");
            String str="我的简历 姓名:西门吹雪 职位:逍遥掌门人 经历:江洋大盗,纵横天下 ";
            student.setResume(str);
            try {
            //将图片转化为输入流
                InputStream in=new FileInputStream("c:\\mypic.jpg");
                byte[] arr=new byte[in.available()];
                //获取输入流中的二进制数据
                in.read(arr);
                //将二进制数据封装到 student 对象的 pic 属性中
                student.setPic(arr);
            } catch (Exception e) {
                // TODO Auto-generated catch block
                e.printStackTrace();
            }
            //将封装好的 student 对象插入数据库中
            int count=studentDao.insertStudent(student);
            if(count>0){
                System.out.println("添加成功!");
            }
        }
    }
```

测试查找学号为1的学生:
```
DEBUG [main] - ==>  Preparing: SELECT * FROM STUDENT where id=?
DEBUG [main] - ==> Parameters: 1(Integer)
DEBUG [main] - <==      Total: 1
我的简历
姓名:张三丰
职位:武当掌门人
经历:武状元,打遍天下无敌手
```

同时观察 C 盘,可发现多了一个文件 mypic.jpg,如图 2.5 所示,从缩略图可知,其正是之前存进去的数据库的图片。

测试插入学号为3的学生:
```
DEBUG [main] - ==>  Preparing: INSERT INTO student VALUES (?, ?, ?, ?)
DEBUG [main] - ==> Parameters: 3(Integer), 西门吹雪(String), [B@156496(byte[]), 我的简历
姓名:西门吹雪 职位:逍遥掌门人 经历:江洋大盗,纵横天下 (String)
DEBUG [main] - <==    Updates: 1
添加成功!
```

图 2.5 从数据库中读取到的图片

观察数据库，如图 2.6 所示，发现图片已存进去了。

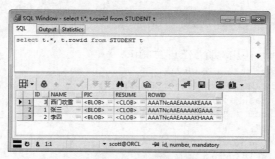

图 2.6 图片已存进数据库

2.12 MyBatis 调用存储过程

项目案例：MySQL 数据库中有一个存储过程，输入班级名称，输出学生人数。在 Java 中调用该存储过程。（项目源码参见配套资源：第 2 章/增删改查/mybatis22）

实现步骤：

（1）在 MySQL 中创建存储过程，代码如下：

```
DELIMITER $$
CREATE
    PROCEDURE 'studentdb'.'sp_student'(IN classnum VARCHAR(10),OUT counts INT)
    BEGIN
SELECT COUNT(*) INTO counts FROM student WHERE classno=classnum;
    END$$
DELIMITER ;
```

（2）在项目 mybatis22 的基础上进行修改。在 StudentMapper 中添加如下代码：

```
<!-- 调用存储过程 -->
    <selectid="getStudentCount"parameterMap="getStudentCountMap"
statementType="CALLABLE">
        CALL sp_student(?,?)
    </select>
    <parameterMap type="java.util.Map" id="getStudentCountMap">
        <parameter property="classnum" mode="IN" jdbcType="VARCHAR"/>
        <parameter property="counts" mode="OUT" jdbcType="INTEGER"/>
    </parameterMap>
```

（3）在接口 IStudentDao 中添加方法 public int getStudentCount(Map map)，在实现类 StudentDao 中实现该方法。代码如下：

```java
//调用存储过程
    @Override
    public int getStudentCount(Map map) {
        SqlSession session = null;
        int num=-1;
        try {
            session = MyBatisUtil.getSession();
            session.selectOne("com.lifeng.dao.IStudentDao.getStudentCount",map);
            num=(int) map.get("counts");
        } catch (Exception e) {
            e.printStackTrace();
        }
        return num;
    }
```

（4）在测试类 TestStudent1 中添加方法进行测试：

```java
public static void getStudentCount(){
    IStudentDao studentDao=new StudentDaoImpl();
    Map map = new HashMap();
    map.put("classnum", "201801");
    map.put("counts", -1);
    int num=studentDao.getStudentCount(map);
    System.out.println(num);
}
```

测试结果如下。
```
DEBUG [main] - ==>  Preparing: CALL sp_student(?,?)
DEBUG [main] - ==> Parameters: 201801(String)
5
```

2.13 MyBatis 逆向工程

在 MyBatis 的项目中，最烦琐的事情就是制作映射文件，还有就是一个数据库表要手动做一个实体类。有一个简便的方法是让 MyBatis 根据数据库连接自动生成映射文件和实体类，这就是 MyBatis 的逆向工程。通常逆向工程单独作为一个项目，把所需的映射文件和实体类自动生成后再复制回真正的项目中，而且自动生成的映射文件往往还要再根据真实的项目需求进行修改。

项目案例：已有 MySQL 数据库，里面有表 Student，通过逆向工程自动生成映射文件和实体类。（项目源码参见配套资源：第 2 章/MyBatis 逆向工程/MybatisGen）

实现步骤：

（1）新建项目 MybatisGen，导入包（见图 2.7）。

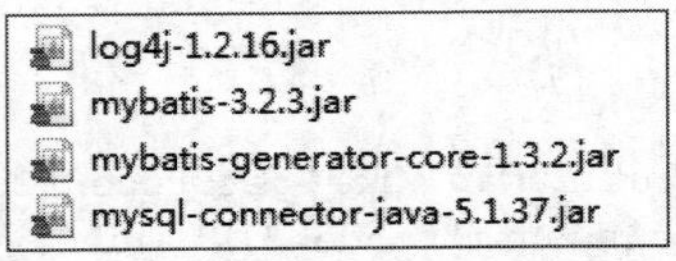

```
log4j-1.2.16.jar
mybatis-3.2.3.jar
mybatis-generator-core-1.3.2.jar
mysql-connector-java-5.1.37.jar
```

图 2.7 逆向工程所需 JAR 包

(2) 在 src 下新建三个包 cn.entity、cn.dao、cn.test 备用。
(3) 在 src 下新建 xml 文件 generatorConfig.xml, 内容如下:

```xml
<?xml version="1.0" encoding="UTF-8"?>
<!DOCTYPE generatorConfiguration
  PUBLIC "-//mybatis.org//DTD MyBatis Generator Configuration 1.0//EN"
  "http://mybatis.org/dtd/mybatis-generator-config_1_0.dtd">
<generatorConfiguration>
    <context id="testTables" targetRuntime="MyBatis3">
        <commentGenerator>
            <!-- 是否去除自动生成的注释 true: 是; false:否 -->
            <property name="suppressAllComments" value="true" />
        </commentGenerator>
        <!--数据库连接的信息: 驱动类、连接地址、用户名、密码 -->
        <jdbcConnection driverClass="com.mysql.jdbc.Driver"
            connectionURL="jdbc:mysql://localhost:3306/studentdb" userId="root"
            password="root">
        </jdbcConnection>
        <!-- 默认 false, 把 JDBC DECIMAL 和 NUMERIC 类型解析为 Integer, 为 true 时把 JDBC
DECIMAL 和 NUMERIC 类型解析为 java.math.BigDecimal -->
        <javaTypeResolver>
            <property name="forceBigDecimals" value="false" />
        </javaTypeResolver>
        <!-- targetProject:生成 PO 类的位置 -->
        <javaModelGenerator targetPackage="cn.entity"
            targetProject=".\src">
            <!-- enableSubPackages:是否让 schema 作为包的后缀 -->
            <property name="enableSubPackages" value="false" />
            <!-- 从数据库返回的值被清理前后的空格 -->
            <property name="trimStrings" value="true" />
        </javaModelGenerator>
        <!-- targetProject:mapper 映射文件生成的位置 -->
        <sqlMapGenerator targetPackage="cn.dao"
            targetProject=".\src">
            <!-- enableSubPackages:是否让 schema 作为包的后缀 -->
            <property name="enableSubPackages" value="false" />
        </sqlMapGenerator>
        <!-- targetPackage: mapper 接口生成的位置 -->
        <javaClientGenerator type="XMLMAPPER"
            targetPackage="cn.dao" targetProject=".\src">
            <!-- enableSubPackages:是否让 schema 作为包的后缀 -->
            <property name="enableSubPackages" value="false" />
        </javaClientGenerator>
        <!-- 指定数据库表 -->
        <!-- <table schema="" tableName="emp"></table> -->
     <!-- 要对那些数据表进行生成操作, 必须要有一个. -->
     <table schema="mybatis" tableName="student" domainObjectName="Student"
        enableCountByExample="false" enableUpdateByExample="false"
        enableDeleteByExample="false" enableSelectByExample="false"
        selectByExampleQueryId="false"> </table>
    </context>
</generatorConfiguration>
```

（4）在包 cn.test 下新建类，代码如下：
```java
package cn.test;
import java.io.File;
import java.io.IOException;
import java.util.ArrayList;
import java.util.List;
import org.mybatis.generator.api.MyBatisGenerator;
import org.mybatis.generator.config.Configuration;
import org.mybatis.generator.config.xml.ConfigurationParser;
import org.mybatis.generator.exception.XMLParserException;
import org.mybatis.generator.internal.DefaultShellCallback;
public class GeneratorSqlmap {
    public void generator() throws Exception{
        List<String> warnings = new ArrayList<String>();
        boolean overwrite = true;
        //指定逆向工程配置文件
        File configFile = new File("src/generatorConfig.xml");
        ConfigurationParser cp = new ConfigurationParser(warnings);
        Configuration config = cp.parseConfiguration(configFile);
        DefaultShellCallback callback = new DefaultShellCallback(overwrite);
        MyBatisGenerator myBatisGenerator = new MyBatisGenerator(config,
                callback, warnings);
        myBatisGenerator.generate(null);
    }
    public static void main(String[] args) throws Exception {
        try {
            GeneratorSqlmap generatorSqlmap = new GeneratorSqlmap();
            generatorSqlmap.generator();
        } catch (Exception e) {
            e.printStackTrace();
        }
    }
}
```

（5）执行上面的类，刷新项目，发现包 cn.entity 下自动生成了一个实体类 Student，包 cn.dao 下自动生成了映射文件和接口文件，如图 2.8 所示。

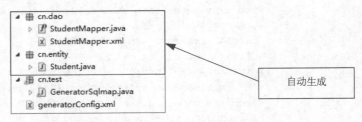

图 2.8　自动生成的文件

（6）把自动生成的实体类、映射文件、接口复制到实际项目中，再根据需要修改即可。

上机练习

题目需求：用第 1 章中的数据库实现查找全部商品；按 ID 号查询商品；添加一个新商品；按 ID 号删除一个商品；按商品名称模糊查询商品；按商品名称、价格、数量等条件组合进行动态查询。

思考题

1. 如何实现模糊查询？
2. 如何实现动态查询？
3. 无 DAO 层实现类如何编程？
4. 如何将视频数据保存到数据库并读取出来？
5. 逆向工程主要解决哪些问题？该如何实现？

第 3 章 多表关联查询

本章目标

- 掌握一对多查询的方法
- 掌握多对一查询的方法
- 掌握多对多查询的方法
- 掌握自连接查询的方法

3.1 一对多查询

一对多关联查询是指在查询一方对象的时候，同时将其所关联的多方对象也都查询出来。

下面以班级 Classes 与学生 Student 间的一对多关系为例进行演示。一个班级有多个学生，一个学生只属于一个班级。数据库 student 表里面有个字段 classno 是外键，对应主键表 Class 的主键 cid。

项目案例：新建项目 mybatis31，查询班级号为 1801 的班级，同时遍历该班的所有学生。（项目源码参见配套资源：第 3 章/一对多/mybatis31）

实现步骤：

（1）复制项目 mybatis22 为 mybatis31，修改配置文件。注意多了几个配置。

别名的配置：

```
<typeAlias alias="Classes" type="com.lifeng.entity.Classes"/>
```

映射文件的配置：

```
<mapper resource="com/lifeng/dao/ClassesMapper.xml"/>
```

其中配置文件如下：

```xml
<?xml version="1.0" encoding="UTF-8"?>
<!DOCTYPE configuration PUBLIC "-//mybatis.org//DTD Config 3.0//EN"
"http://mybatis.org/dtd/mybatis-3-config.dtd">
<configuration>
    <!-- 注册属性文件 -->
    <properties resource="jdbc.properties"/>
    <typeAliases>
        <typeAlias alias="Student" type="com.lifeng.entity.Student"/>
        <typeAlias alias="Classes" type="com.lifeng.entity.Classes"/>
    </typeAliases>
    <environments default="development">
        <environment id="development">
            <transactionManager type="JDBC"/>
            <dataSource type="POOLED">
                <property name="driver" value="${jdbc.driver}"/>
                <property name="url" value="${jdbc.url}"/>
                <property name="username" value="${jdbc.username}"/>
                <property name="password" value="${jdbc.password}"/>
            </dataSource>
        </environment>
    </environments>
    <mappers>
        <mapper resource="com/lifeng/dao/StudentMapper.xml"/>
        <mapper resource="com/lifeng/dao/ClassesMapper.xml"/>
    </mappers>
</configuration>
```

（2）修改数据库。

修改表 student，添加一列 classno 类型 varchar(10)，可为空，该列添加值 201801：

```
ALTER TABLE student ADD COLUMN classno VARCHAR(10);
UPDATE student SET classno='201801';
```

新建数据表 classes 并添加数据：

```
create table classes (
    cid varchar (30),
```

```
        cname varchar (60)
);
insert into classes (cid, cname) values('201801','计算机软件1班');
insert into classes (cid, cname) values('201802','计算机软件2班');
```

（3）创建实体类 Student 和 Classes。

Student 类如下：

```
package com.lifeng.entity;
public class Student {
    private String sid;
    private String sname;
    private String sex;
    private int age;
    //所在班级
    private Classes classes;
    //省略getter,setter方法
    public Student(){}
    public Student(String sname,String sex,int age){
        this.sname=sname;
        this.sex=sex;
        this.age=age;
    }
    public Student(String sid,String sname,String sex,int age){
        this.sid=sid;
        this.sname=sname;
        this.sex=sex;
        this.age=age;
    }
    public void show(){
        System.out.println("学生编号:"+sid+" 学生姓名:"+sname+" 学生性别"+sex+" 学生年龄:"+age);
    }
}
```

Classes 类如下：

```
package com.lifeng.entity;
import java.util.List;
public class Classes {
    private String cid;
    private String cname;
    private List<Student> students;
    //省略getter,setter方法
    public String toString(){
        return "班级编号:"+getCid()+",班级名称:"+getCname();
    }
}
```

（4）创建 DAO 接口。

```
public interface IClassesDao {
    public Classes findClassesById(String id);
}
```

（5）创建映射文件 ClassesMapper.xml，有以下两种方式。

① 方式一：多表连接查询方式。

这种方式只用到1条 SQL 语句，代码如下所示：

```xml
<mapper namespace="com.lifeng.dao.IClassesDao">
    <resultMap id="classesResultMap" type="com.lifeng.entity.Classes">
        <id property="cid" column="cid" />
        <result property="cname" column="cname" />
        <!-- 关联属性的映射关系 -->
        <collection property="students" ofType="Student">
            <id property="sid" column="id" />
            <result property="sname" column="studentname" />
            <result property="sex" column="gender" />
            <result property="age" column="age" />
        </collection>
    </resultMap>
    <!-- 多表连接查询 -->
    <select id="findClassesById" parameterType="String" resultMap="classesResultMap">
        select cid,cname,id,studentname,gender,age from classes,student
        where classes.cid=student.classno
        and classes.cid=#{cid}
    </select>
</mapper>
```

【注意】即使字段名与属性名相同，在<resultMap/>中也要写出它们的映射关系。因为框架是依据这个<resultMap/>封装对象的。

另外，在映射文件中使用<collection/>标签体现出两个实体对象间的关联关系。其两个属性的解释如下。

- property：指定关联属性，即 Class 类中的集合属性 students。
- ofType：集合属性的泛型类型，即 Student。

② 方式二：多表单独查询方式。

多表连接查询方式是将多张表进行连接，连为一张表后再进行查询，其查询的本质是一张表。而多表单独查询方式是多张表各自查询各自的相关内容，需要多张表的联合数据，再将主表的查询结果联合其他表的查询结果，封装为一个对象。

多表查询是可以跨越多个映射文件的，即可以跨越多个 namespace。在使用其他 namespace 查询时，添加上其所在的 namespace 即可。这种方式要用到 2 条 SQL 语句，代码如下所示：

```xml
<resultMap id="studentResultMap" type="com.lifeng.entity.Student">
    <id property="sid" column="id" />
    <result property="sname" column="studentname" />
    <result property="sex" column="gender" />
    <result property="age" column="age" />
</resultMap>
resultMap id="classesResultMap" type="com.lifeng.entity.Classes">
    <id property="cid" column="cid" />
    <result property="cname" column="cname" />
    <!-- 关联属性的映射关系 -->
    <!-- 集合的数据来自指定的 select 查询,该 select 查询的动态参数来自 column 指定的字段值 -->
<collection property="students" ofType="Student"
        select="selectStudentsByClasses" column="cid"/>
</resultMap>
<!-- 多表单独查询,查多方的表 -->
<select id="selectStudentsByClasses" resultMap="studentResultMap">
    select * from student where calssno=#{cid}
</select>
```

```xml
<!-- 多表单独查询,查一方的表 -->    1)
<select id="findClassesById" parameterType="String" resultMap="classesResultMap">
    select cid,cname from classes
    where cid=#{cid}
</select>
```

关联属性<collection/>的数据来自另一个查询< selectStudentsByClasses />，而该查询< selectStudentsByClasses />的动态参数 calssno=#{cid}的值来自< findClassesById />的查询结果字段 cid。

以上箭头所示的流程如下。

① 先执行 findClassesById 查询主表 classes，查询出各个列的值，包括 cid。

② 将查询主表出来的各个列值传递给<resultMap id="classesResultMap" >进行结果映射。其中 cid 的值还要传给<collection>节点作为实参。

③ 执行 selectStudentsByClasses 查询，以 cid 作为实参，查询出结果，结果封装为<resultMap id="studentResultMap">。

（6）实现类 ClassesDaoImpl，代码如下：

```java
@Override
public Classes findClassesById(String id) {
    Classes classes = new Classes();
    try {
        session = MyBatisUtil.getSession();
        classes = session.selectOne("com.lifeng.dao.IClassesDao.findClassesById", id);
    } catch (Exception e) {
        e.printStackTrace();
    }finally {
        session.close();
    }
    return classes;
}
```

（7）在测试类中添加如下方法：

```java
public static void main(String[] args) {
    findClassesById();
}
public static void findClassesById(){
    IClassesDao classesDao=new ClassesDaoImpl();
    Classes mycalss=classesDao.findClassesById("201801");
    System.out.println("班级编号:"+mycalss.getCid()+",班级名称:"+mycalss.getCname()+",班级学生:");
    List<Student> slist=mycalss.getStudents();
    for(int i=0;i<slist.size();i++){
        slist.get(i).show();
    }
}
```

方式一：多表连接查询方式测试结果。

```
DEBUG [main] - ==>  Preparing: select cid,cname,id,studentname,gender,age from classes,student where classes.cid=student.classno and classes.cid=?
DEBUG [main] - ==> Parameters: 201801(String)        只有一条SQL语句

DEBUG [main] - <==      Total: 4
班级编号:201801,班级名称:计算机软件1班,班级学生:
```

班级编号:201801,班级名称:计算机软件1班,班级学生:
学生编号:1 学生姓名:张飞 学生性别男 学生年龄:18
学生编号:2 学生姓名:李白 学生性别男 学生年龄:20
学生编号:3 学生姓名:张无忌 学生性别男 学生年龄:19
学生编号:4 学生姓名:赵敏 学生性别女 学生年龄:18

可以发现,它发出的 SQL 语句是多表联查。

方式二:多表单独查询方式测试结果。

```
DEBUG [main] - ==>  Preparing: select cid,cname from classes where cid=?
DEBUG [main] - ==> Parameters: 201801(String)
DEBUG [main] - ====>  Preparing: select * from student where classno=?
DEBUG [main] - ====> Parameters: 201801(String)
DEBUG [main] - <====      Total: 4
DEBUG [main] - <==       Total: 1
```
两条 SQL 语句

班级编号:201801,班级名称:计算机软件1班,班级学生:
学生编号:1 学生姓名:张飞 学生性别男 学生年龄:18
学生编号:2 学生姓名:李白 学生性别男 学生年龄:20
学生编号:3 学生姓名:张无忌 学生性别男 学生年龄:19
学生编号:4 学生姓名:赵敏 学生性别女 学生年龄:17

可以发现,其 SQL 语句是两条,即各查各的,共用同一个参数。第 1 条先查一方的表,第 2 条再查多方的表。

3.2 多对一关联查询

多对一关联查询是指在查询多方对象的时候,同时将其所关联的一方对象也查询出来。

由于在查询多方对象时也是一个一个地查询,所以多对一关联查询其实就是一对一关联查询,即一对一关联查询的实现方式与多对一的实现方式是相同的。配置多对一关联的重点在于"多方"的映射文件要有<association>属性关联"一方"。

项目案例:查询学号为 1 的学生,同时获取他所在班级的完整信息。(项目源码参见配套资源:第 3 章/多对一/mybatis32)

实现步骤:

(1)复制项目 mybatis31 为 mybatis32,在接口 IStudentDao 中添加方法 searchStudentsById(String id)如下:

```
public Student searchStudentsById(int id);
```

(2)配置映射文件 ClassesMapper.xml,内容如下。
方式一:多表联合查询。

```xml
<resultMap id="StudentMapper" type="Student">
    <id property="sid" column="id" />
    <result property="sname" column="studentname" />
    <result property="sex" column="gender" />
    <result property="age" column="age" />
    <!-- 关联属性 -->
    <association property="classes" javaType="Classes">
        <id property="cid" column="cid" />
        <result property="cname" column="cname" />
```

```xml
        </association>
    </resultMap>
    <!-- 多表连接查询 -->
    <select id="searchStudentsById" parameterType="int" resultMap="StudentMapper">
        select cid,cname,id,studentname,gender,age
        from classes,student
        where classes.cid=student.classno
        and student.id=#{id}
    </select>
```

方式二：多表单独查询。

```xml
<resultMap id="studentResultMap2" type="com.lifeng.entity.Student">
    <id property="sid" column="id" />
    <result property="sname" column="studentname" />
    <result property="sex" column="gender" />
    <result property="age" column="age" />
    <!-- 关联属性 -->
    <association property="classes" javaType="Classes"
     select="findClassesById" column="classno"/>
</resultMap>
<select id="searchStudentsById" resultMap="studentResultMap2">
    select id,studentname,gender,age,classno from student where id=#{id}
</select>
<select id="findClassesById" parameterType="String" resultType="Classes">
    select cid,cname from classes
    where cid=#{cid}
</select>
```

箭头所指的意思如下。

① 将<select id="searchStudentsById">中查询出来的 classno 传递给<association>中的 column="classno"。

② 再把 classno 的值作为实参，传递给<select id="findClassesById">中的形参#{cid}。

（3）在实现类 StudentDaoImpl 中添加如下方法：

```java
@Override
public Student searchStudentsById(int id) {
    Student student = new Student();
    try {
        session = MyBatisUtil.getSession();
        student = session.selectOne("com.lifeng.dao.IClassesDao.searchStudentsById",id);
    } catch (Exception e) {
        e.printStackTrace();
    }finally {
        session.close();
    }
    return student;
}
```

（4）在测试类中添加如下方法：

```java
public static void main(String[] args) {
    searchStudentsById();
}
public static void searchStudentsById(){
    IStudentDao studentDao=new StudentDaoImpl();
    Student student=studentDao.searchStudentsById(1);
```

```
            student.show();
            System.out.println("所在班级:");
            Classes classes=student.getClasses();
            System.out.println(classes.toString());
        }
```

测试方式一如下:
```
DEBUG [main] - ==>  Preparing: select cid,cname,id,studentname,gender,age from classes,student where classes.cid=student.classno and student.id=?
DEBUG [main] - ==> Parameters: 1(Integer)
DEBUG [main] - <==      Total: 1
```
学生编号:1 学生姓名:张飞 学生性别男 学生年龄:18
所在班级:
班级编号:201801,班级名称:计算机软件1班

可以发现，它发出的 SQL 语句是多表联查。

测试方式二如下:
```
DEBUG [main] - ==>  Preparing: select id,studentname,gender,age,classno from student where id=?
DEBUG [main] - ==> Parameters: 1(Integer)
DEBUG [main] - ====>  Preparing: select cid,cname from classes where cid=?
DEBUG [main] - ====> Parameters: 201801(String)
DEBUG [main] - <====      Total: 1
DEBUG [main] - <==      Total: 1
```
学生编号:1 学生姓名:张飞 学生性别男 学生年龄:18
所在班级:
班级编号:201801,班级名称:计算机软件1班

可以发现，其 SQL 语句是两条，即各查各的，共用同一个参数。

3.3 自连接

自连接的查询可以用一对多来处理，也可以用多对一来处理。例如，员工表，每个员工都有一个上司，但上司同时也是员工表的一条记录，这种情况可用自连接查询出每个员工对应的上司信息，也可以查出每个上司有哪些下属员工。

3.3.1 使用多对一的方式实现自连接

项目案例：查询员工的信息及对应的上司信息。（项目源码参见配套资源：第3章/自连接多对一方式/mybatis33）

思路分析：可将员工当作多方，上司当作一方。

实现步骤：

（1）修改数据库。

添加一个表 employee 并插入测试数据，具体如下：
```
create table employee (
    empid double ,
    empname varchar (60),
    job varchar (60),
    leader double
```

```
);
insert into employee (empid, empname, job, leader) values('1','jack','clerk','3');
insert into employee (empid, empname, job, leader) values('2','mike','salesman','3');
insert into employee (empid, empname, job, leader) values('3','john','manager','4');
insert into employee (empid, empname, job, leader) values('4','smith','president',NULL);
insert into employee (empid, empname, job, leader) values('5','rose','salesman','3');
```

（2）创建实体类 Employee，代码如下：

```
package com.lifeng.entity;
public class Employee {
    private int empid;
    private String empname;
    private String job;
    private Employee leader;
    //省略 setter,getter 方法
    public String toString(){
        return "员工编号:"+getEmpid()+",员工姓名:"+getEmpname()+",员工职位:"+getJob();
    }
}
```

可以发现，里面存在着嵌套，Employee 里面的一个属性 leader 本身就是 Employee 类型。

（3）创建 IEmployeeDao 接口，添加方法 findEmployeeAndLeaderById(int id)如下：

```
public Employee findEmployeeAndLeaderById(int id);
```

（4）修改 Mybatis-config.xml 配置文件，代码如下：

```xml
<?xml version="1.0" encoding="UTF-8"?>
<!DOCTYPE configuration PUBLIC "-//mybatis.org//DTD Config 3.0//EN"
"http://mybatis.org/dtd/mybatis-3-config.dtd">
<configuration>
    <!-- 注册属性文件 -->
    <properties resource="jdbc.properties"/>
    <typeAliases>
        <typeAlias alias="Student" type="com.lifeng.entity.Student"/>
        <typeAlias alias="Classes" type="com.lifeng.entity.Classes"/>
        <typeAlias alias="Employee" type="com.lifeng.entity.Employee"/>
    </typeAliases>
    <environments default="development">
        <environment id="development">
            <transactionManager type="JDBC"/>
            <dataSource type="POOLED">
                <property name="driver" value="${jdbc.driver}"/>
                <property name="url" value="${jdbc.url}"/>
                <property name="username" value="${jdbc.username}"/>
                <property name="password" value="${jdbc.password}"/>
            </dataSource>
        </environment>
    </environments>
    <mappers>
        <mapper resource="com/lifeng/dao/StudentMapper.xml"/>
        <mapper resource="com/lifeng/dao/ClassesMapper.xml"/>
        <mapper resource="com/lifeng/dao/EmployeeMapper.xml"/>
    </mappers>
</configuration>
```

（5）创建映射文件 EmployeeMapper，代码如下：

```xml
<?xml version="1.0" encoding="utf-8" ?>
```

```xml
<!DOCTYPE mapper PUBLIC "-//mybatis.org//DTD Mapper 3.0//EN"
"http://mybatis.org/dtd/mybatis-3-mapper.dtd">
<mapper namespace="com.lifeng.dao.IEmployeeDao">
    <resultMap id="employeeMap" type="com.lifeng.entity.Employee">
        <id property="empid" column="empid" />
        <result property="empname" column="empname" />
        <result property="job" column="job" />
        <association property="leader" javaType="Employee"
        select="findEmployeeAndLeaderById" column="leader"/>
    </resultMap>
    <selectid="findEmployeeAndLeaderById"parameterType="int" resultMap="employeeMap">
        select * from employee where empid=#{empid}
    </select>
</mapper>
```

（6）接口实现类 EmployeeDaoImpl.java 如下：

```java
@Override
public Employee findEmployeeAndLeaderById(int id) {
    SqlSession session = null;
    Employee employee=new Employee();
    try {
        session = MyBatisUtil.getSession();
        employee = session.selectOne("com.lifeng.dao.IEmployeeDao.findEmployeeAndLeaderById",id);
    } catch (Exception e) {
        e.printStackTrace();
    }finally {
        session.close();
    }
    return employee;
}
```

（7）测试类 TestEmployee1.java 如下：

```java
public static void main(String[] args) {
    findEmployeeAndLeaderById();
}
public static void findEmployeeAndLeaderById(){
    Scanner input=new Scanner(System.in);
    System.out.print("请输入要查询的员工编号:");
    int empid=input.nextInt();
    IEmployeeDao eDao=new EmployeeDaoImpl();
    Employee employee=eDao.findEmployeeAndLeaderById(empid);
    Employee leader=employee.getLeader();
    System.out.println(employee.toString());
    System.out.println("他的上司是:"+leader.toString());
}
```

测试结果一：查询普通职员。

```
请输入要查询的员工编号:1
DEBUG [main] - ==> Preparing: select * from employee where empid=?
DEBUG [main] - ==> Parameters: 1(Integer)
DEBUG [main] - ====> Preparing: select * from employee where empid=?
DEBUG [main] - ====> Parameters: 3(Integer)
DEBUG [main] - ======> Preparing: select * from employee where empid=?
DEBUG [main] - ======> Parameters: 4(Integer)
DEBUG [main] - <======      Total: 1
```

```
DEBUG [main] - <====      Total: 1
DEBUG [main] - <==       Total: 1
```
员工编号:1,员工姓名:jack,员工职位:clerk
他的上司是:员工编号:3,员工姓名:john,员工职位:manager

从上面的 SQL 语句发现，出现了 3 条 SQL 语句，这个查询存在嵌套，先查员工1，然后查他的直接上司3，再查上司的上司4。这种情况不影响什么，甚至可以实现直接输出上司的上司，但要注意输出语句不要出现递归，即输出语句不要出现输出上司。

要同时查上司的上司，只需要在上面的测试类中多加一条语句。
```
System.out.println("他的上司的上司是:"+leader.getLeader().toString());
```

测试结果二：查询经理，经理的上司是 CEO。
```
请输入要查询的员工编号:3
DEBUG [main] - ==> Preparing: select * from employee where empid=?
DEBUG [main] - ==> Parameters: 3(Integer)
DEBUG [main] - ====> Preparing: select * from employee where empid=?
DEBUG [main] - ====> Parameters: 4(Integer)
DEBUG [main] - <====      Total: 1
DEBUG [main] - <==       Total: 1
```
员工编号:3,员工姓名:john,员工职位:manager
他的上司是:员工编号:4,员工姓名:smith,员工职位:CEO

3.3.2 使用一对多方式实现自连接

项目案例：查询某位领导及其直接下属员工。（项目源码参见本书配套资源：第 3 章/自连接一对多方式/mybatis34）

思路分析：可用一对多的方式来实现，员工（领导）当作一方，员工（下属）当作多方。

实现步骤：

（1）复制 mybatis33 为 mybatis34，修改实体类，添加属性，具体如下：
```
private List<Employee> employees;
public List<Employee> getEmployees() {
    return employees;
}
public void setEmployees(List<Employee> employees) {
    this.employees = employees;
}
```
（2）创建 IEmployeeDao 接口，添加方法如下：
```
public List<Employee> findLeaderAndEmployeesById(int id);
```
（3）在映射文件 EmployeeMapper 中添加如下内容：
```
<resultMap id="empResultMap2" type="com.lifeng.entity.Employee">
    <id property="empid" column="empid" />
    <result property="empname" column="empname" />
    <result property="job" column="job" />
    <!-- 关联属性的映射关系
集合的数据来自指定的 select 查询,该 select 查询的动态参数来自 column 指定的字段值 -->
    <collection property="employees" ofType="Employee"
        select="selectEmployeesByLeader" column="empid"/>
</resultMap>
<select id="selectEmployeesByLeader" resultType="Employee">
```

```xml
        select * from employee where leader=#{empid}
    </select>
    <select id="findLeaderAndEmployeesById" parameterType="int" resultMap="empResultMap2">
        select * from employee where empid=#{empid}
    </select>
```

（4）在接口实现类 EmployeeDaoImpl.java 中添加方法 findLeaderAndEmployeesById(int id)如下：

```java
@Override
public Employee findLeaderAndEmployeesById(int id) {
    SqlSession session = null;
    Employee employee=new Employee();
    try {
        session = MyBatisUtil.getSession();
        employee=session.selectOne("com.lifeng.dao.IEmployeeDao.findLeaderAndEmployeesById",id);
    } catch (Exception e) {
        e.printStackTrace();
    }
    return employee;
}
```

（5）测试类如下：

```java
public static void main(String[] args) {
    findLeaderAndEmployeesById();
}
public static void findLeaderAndEmployeesById(){
    Scanner input=new Scanner(System.in);
    System.out.print("请输入要查询的员工编号:");
    int empid=input.nextInt();
    EmployeeDaoImpl eDao=new EmployeeDaoImpl();
    Employee leader=eDao.findLeaderAndEmployeesById(empid);
    List<Employee> employees=leader.getEmployees();
    System.out.println(leader.toString());
    System.out.println("他的直接下属有:");
    for(Employee emp:employees){
        System.out.println(emp.toString());
    }
}
```

测试结果：查询经理。

```
请输入要查询的员工编号:3
DEBUG [main] - ==>  Preparing: select * from employee where empid=?
DEBUG [main] - ==> Parameters: 3(Integer)
DEBUG [main] - ====>  Preparing: select * from employee where leader=?
DEBUG [main] - ====> Parameters: 3(Integer)
DEBUG [main] - <====      Total: 3
DEBUG [main] - <==      Total: 1
员工编号:3,员工姓名:john,员工职位:manager
他的直接下属有:
员工编号:1,员工姓名:jack,员工职位:clerk
员工编号:2,员工姓名:mike,员工职位:salesman
员工编号:5,员工姓名:rose,员工职位:salesman
```

测试结果：查询 CEO。

```
请输入要查询的员工编号:4
DEBUG [main] - ==>  Preparing: select * from employee where empid=?
DEBUG [main] - ==> Parameters: 4(Integer)
DEBUG [main] - ====>  Preparing: select * from employee where leader=?
DEBUG [main] - ====> Parameters: 4(Integer)
DEBUG [main] - <=====      Total: 1
DEBUG [main] - <==      Total: 1
员工编号:4,员工姓名:smith,员工职位:CEO
他的直接下属有:
员工编号:3,员工姓名:john,员工职位:manager
```

3.4 多对多查询

原理：多对多可以拆分成两个一对多来处理，需要一个中间表，各自与中间表实现一对多的关系。

项目案例：一个学生可以选修多门课程，一门课程可以给多个学生选修，课程与学生之间是典型的多对多关系。如何查询一个学生的信息，同时查出他的所有选修课；另外，如何实现查询一门课程信息，同时查出所有选修了该课程的学生信息。（项目源码参见本书配套资源：第 3 章/多对多/mybatis35）

思路分析：多对多关系需要第三张表来体现，数据库中除了课程表、学生表，还需要学生课程表。

实现步骤：

（1）修改数据库，代码如下：

```
create table course (
    courseid double ,
    coursename varchar (90)
);
insert into course (courseid, coursename) values('1','java');
insert into course (courseid, coursename) values('2','android');
insert into course (courseid, coursename) values('3','PHP');
create table studentcourse (
    id double ,
    studentid double ,
    courseid double
);
insert into studentcourse (id, studentid, courseid) values('1','1','1');
insert into studentcourse (id, studentid, courseid) values('2','1','2');
insert into studentcourse (id, studentid, courseid) values('3','2','1');
insert into studentcourse (id, studentid, courseid) values('4','2','2');
insert into studentcourse (id, studentid, courseid) values('5','3','1');
insert into studentcourse (id, studentid, courseid) values('6','3','2');
insert into studentcourse (id, studentid, courseid) values('7','1','3');
```

（2）复制 mybatis34 为 mybatis35，新增实体类 Course 和修改实体类 Student。

Course 类如下：

```
package com.lifeng.entity;
import java.util.List;
public class Course {
    private int courseid;
    private String coursename;
    private List<Student> students;
```

```java
    //省略 setter,getter 方法
    public String toString(){
        return "课程编号:"+getCourseid()+",课程名称:"+getCoursename();
    }
}
```

Student 类如下，添加一个属性 courses 和 getter、setter 方法。
```java
private List<Course> courses;
public List<Course> getCourses() {
    return courses;
}
public void setCourses(List<Course> courses) {
    this.courses = courses;
}
```

（3）修改 mybatis-config.xml 配置文件，代码如下：
```xml
<?xml version="1.0" encoding="UTF-8"?>
<!DOCTYPE configuration PUBLIC "-//mybatis.org//DTD Config 3.0//EN"
"http://mybatis.org/dtd/mybatis-3-config.dtd">
<configuration>
    <!-- 注册属性文件 -->
    <properties resource="jdbc.properties"/>
    <typeAliases>
        <typeAlias alias="Student" type="com.lifeng.entity.Student"/>
        <typeAlias alias="Classes" type="com.lifeng.entity.Classes"/>
        <typeAlias alias=" Course" type="com.lifeng.entity.Course"/>
    </typeAliases>
    <environments default="development">
        <environment id="development">
            <transactionManager type="JDBC"/>
            <dataSource type="POOLED">
                <property name="driver" value="${jdbc.driver}"/>
                <property name="url" value="${jdbc.url}"/>
                <property name="username" value="${jdbc.username}"/>
                <property name="password" value="${jdbc.password}"/>
            </dataSource>
        </environment>
    </environments>
    <mappers>
        <mapper resource="com/lifeng/dao/StudentMapper.xml"/>
        <mapper resource="com/lifeng/dao/ClassesMapper.xml"/>
        <mapper resource="com/lifeng/dao/CourseMapper.xml"/>
    </mappers>
</configuration>
```

（4）在接口 IStudentDao 中添加一个方法 searchStudentById(int id)如下：
```java
public Student searchStudentById(int id);
```

（5）新建接口 ICourseDao，添加一个方法 searchCourseById(int id)如下：
```java
public Course searchCourseById(int id);
```

（6）配置 Mapper 映射。

StudentMapper 的配置如下：
```xml
<?xml version="1.0" encoding="utf-8" ?>
<!DOCTYPE mapper PUBLIC "-//mybatis.org//DTD Mapper 3.0//EN"
"http://mybatis.org/dtd/mybatis-3-mapper.dtd">
```

```xml
<mapper namespace="com.lifeng.dao.IStudentDao">
    <resultMap id="studentMap2" type="com.lifeng.entity.Student">
        <id property="sid" column="id" />
        <result property="sname" column="studentname" />
        <result property="sex" column="gender" />
        <result property="age" column="age" />
        <!-- 关联属性的映射关系 -->
        <collection property="courses" ofType="Course">
            <id property="courseid" column="courseid" />
            <result property="coursename" column="coursename" />
        </collection>
    </resultMap>
    <!-- 多表连接查询 -->
    <select id="searchStudentById" parameterType="int" resultMap="studentMap2">
        select student.id,studentname,gender,age,course.courseid,coursename from
     course,student,studentcourse
        where course.courseid=studentcourse.courseid
        and student.id=studentcourse.studentid and student.id=#{id}
    </select>
</mapper>
```

很容易看出上面的配置是一对多的情形。

CourseMapper 的配置如下：

```xml
<?xml version="1.0" encoding="utf-8" ?>
<!DOCTYPE mapper PUBLIC "-//mybatis.org//DTD Mapper 3.0//EN"
 "http://mybatis.org/dtd/mybatis-3-mapper.dtd">
<mapper namespace="com.lifeng.dao.ICourseDao">
    <resultMap id="courseMap" type="com.lifeng.entity.Course">
        <id property="courseid" column="courseid" />
        <result property="coursename" column="coursename" />
        <!-- 关联属性的映射关系 -->
        <collection property="students" ofType="Student">
            <id property="sid" column="id" />
            <result property="sname" column="studentname" />
            <result property="sex" column="gender" />
            <result property="age" column="age" />
        </collection>
    </resultMap>
    <!-- 多表连接查询 -->
    <select id="searchCourseById" parameterType="int" resultMap="courseMap">
        selectid,studentname,gender,age,courseid,coursenamefrom
course,student,studentcourse
        where course.courseid=studentcourse.courseid
        and student.id=studentcourse.studentid and course.courseid=#{courseid}
    </select>
</mapper>
```

这个配置也是一对多的情形，所以多对多可以拆分成两个一对多。

（7）实现类如下。

StudentDaoImpl 类如下：

```java
@Override
public Student searchStudentById(int id) {
    Student student = new Student();
    try {
```

```
                session = MyBatisUtil.getSession();
                student = session.selectOne("com.lifeng.dao.IStudentDao.searchStudentById",id);
            } catch (Exception e) {
                e.printStackTrace();
            }finally {
                session.close();
            }
            return student;
        }
```

CourseDaoImpl 类如下：
```
@Override
public Course searchCourseById(int id) {
    Course course = new Course();
    try {
        session = MyBatisUtil.getSession();
        course = session.selectOne("com.lifeng.dao.ICourseDao.searchCourseById",id);
    } catch (Exception e) {
        e.printStackTrace();
    }finally {
        session.close();
    }
    return course;
}
```

（8）测试类。

TestStudent1.java 中添加方法 **searchStudentById**，查询 1 号学生信息及其选修课信息。
```
public static void main(String[] args) {
    searchStudentById();
}
public static void searchStudentById(){
    IStudentDao studentDao=new StudentDaoImpl();
    Student student=studentDao.searchStudentById(1);
    student.show();
    System.out.println("-----该生选修了以下课程:-----------");
    List<Course> courses=student.getCourses();
    for(Course course:courses){
        System.out.println(course.toString());
    }
}
```

测试结果如下：
```
DEBUG [main]-==>Preparing:select student.id,studentname,gender,age,course.courseid,coursename from course,student,studentcourse where course.courseid=studentcourse.courseid and student.id=studentcourse.studentid and student.id=?
DEBUG [main] - ==> Parameters: 1(Integer)
DEBUG [main] - <==       Total: 3
学生编号:1 学生姓名:张飞 学生性别男 学生年龄:18
-----该生选修了以下课程:-----------
课程编号:1,课程名称:Java
课程编号:2,课程名称:Android
课程编号:3,课程名称:PHP
```

在 **TestCourse1.java** 中添加方法 **searchCourseById**，查询 1 号课的信息及选修的学生信息。
```
public static void main(String[] args) {
```

```
            searchCourseById();
    }
    public static void searchCourseById(){
            ICourseDao courseDao=new CourseDaoImpl();
            Course course=courseDao.searchCourseById(1);
            System.out.println(course.toString());
            System.out.println("-------该课程有以下学生选修:------");
            List<Student> students=course.getStudents();
            for(Student student:students){
                student.show();
            }
    }
```

测试结果如下:

DEBUG[main]-==>Preparing:selectstudent.id,studentname,gender,age,course.courseid,coursename from course,student,studentcoursewherecourse.courseid=studentcourse.courseidand student.id=studentcourse.studentid and course.courseid=?
DEBUG [main] - ==> Parameters: 1(Integer)
DEBUG [main] - <== Total: 3
课程编号:1,课程名称:Java
-------该课程有以下学生选修:------
学生编号:1 学生姓名:张飞 学生性别男 学生年龄:18
学生编号:2 学生姓名:李白 学生性别男 学生年龄:20
学生编号:3 学生姓名:张无忌 学生性别男 学生年龄:19

上机练习

1. 修改商品表,再添加一列:商品类别。(商品表的内容见表 3.1)

表 3.1 商品表

商品编号 id	商品名称 goodsname	商品单价 price	商品数量 quantity	商品类别 typeid
1	电视机	5000	100	1
2	电冰箱	4000	200	2
3	空调	3000	300	2
4	洗衣机	3500	400	2

再建一个商品类别表,如表 3.2 所示。

表 3.2 商品类别表

商品类别编号 tid	商品类别名称
1	黑色家电
2	白色家电

要求:

(1)查询商品表时同时输出其类别名称(多对一)。

(2)查询商品类别表时同时输出其商品集合(一对多)。

2. 创建一个订单表,订单明细表、商品表与订单表的关系是多对多的关系,见表 3.3 和表 3.4。

表 3.3 订单表

订单编号 orderid	订购单位 buyer	交货日期
1	公司 A	2019-10-1
2	公司 B	2019-10-2
3	公司 C	2019-10-3
4	公司 D	2018-10-4

表 3.4 订单明细表

编号 orderitemid	订单编号 orderid	商品编号 goodsid
1	1	1
2	1	2
3	2	2
4	2	3

要求：
（1）查询1号订单有哪些商品。
（2）查询在哪些订单里有2号商品。

思考题

1. 一对多怎么实现？
2. 多对一怎么实现？
3. 多对多怎么实现？
4. 自连接怎么实现？

第 4 章　Spring 入门

本章目标

- 理解 Spring 的体系结构
- 理解 Spring 的思想
- 理解 IoC 的原理
- 掌握搭建 Spring 开发环境的方法

4.1 Spring 概述

Spring 是为了解决企业应用开发的复杂性而创建的一个轻量级的 Java 开发框架。Spring 的核心是控制反转（Inversion of Control，IoC）和面向切面编程（Aspect Oriented Programming，AOP）。传统的 Java 程序，类与类之间存在较强的依赖关系，增加了程序开发的难度，开发某一个类的时候还要考虑对另一个类的影响，一个类的修改往往导致另一个类不得不跟着修改，程序可维护性和可拓展性变差。使用 Spring 可以降低代码间的依赖程度（即耦合度），为代码"解耦"，提高程序的可拓展性、可复用性和可维护性，使主业务专注于自身的开发。

Spring 降低耦合度的方式有 IoC 与 AOP 两种。IoC 的作用是使得主业务在相互调用过程中，不用再自己维护关系，即无须自己创建要使用的对象，而是由 Spring 容器统一管理，自动"注入"。AOP 技术不用再由程序员用硬编码的方式将系统级服务"混杂"到主业务的逻辑中，而是由 Spring 容器统一完成"织入"。

Spring 还提供对其他框架的支持，如支持 MyBatis、Hibernate、Struts 等框架，简化这些框架的应用，通过整合使它们高效地协同工作。

4.1.1 Spring 的体系结构

Spring 包括 7 大功能模块，分别是 Core、AOP、ORM、DAO、MVC、Web 和 Context。Spring 体系架构如图 4.1 所示。

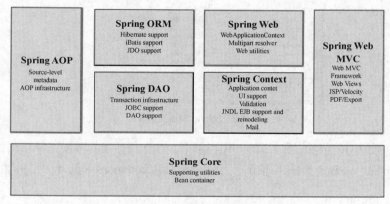

图 4.1 Spring 体系架构图

1. Core 模块

Core 模块是 Spring 的核心类库，提供 IoC 功能。

2. AOP 模块

AOP 模块是 Spring 的 AOP 库，提供了 AOP 拦截器机制。

3. ORM 模块

Spring 的 ORM 模块提供对常用的 ORM 框架的管理和辅助支持，支持流行的 Hibernate、iBatis 等框架，但 Spring 本身并不对 ORM 进行实现，仅对常见的 ORM 框架进行封装，并对其进行管理。

4. DAO 模块

Spring 提供对 JDBC 的支持，对 JDBC 进行封装，允许 JDBC 使用 Spring 资源，并能统一管理 JDBC 事务。

5. MVC 模块

MVC 模块为 Spring 提供了一套轻量级的 MVC 实现，在 Spring 的开发中，既可以用 Struts2 也可以用 Spring 自己的 MVC 框架，相对于 Struts2，Spring 自己的 MVC 框架更加简洁和方便。

6. Web 模块

Web 模块提供对常见框架（如 Struts2、JSF 等）的支持，让 Spring 能够管理这些框架，将 Spring 的资源注入框架，也能在这些框架的前后插入拦截器。

7. Context 模块

Context 模块提供 Bean 访问方式。其他程序可以通过 Context 访问 Spring 的 Bean 资源，相当于资源注入。

4.1.2 Spring 的开发环境

安装 Spring 的开发环境需要先到官网下载 Spring 的 JAR 包，本书用的是 Spring4.3.4 版本，下载 spring-framework-4.3.4.RELEASE-dist.zip 文件，解压后其结构如图 4.2 所示。

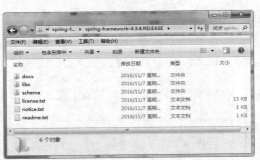

图 4.2　Spring 解压文件夹

其中，Spring 的 JAR 包就在图 4.2 的 libs 目录下，在项目开发的时候，要将用到的 JAR 包复制到项目的 WebContent/WEB-INF/lib 目录下。此外还要用到 commons-logging-1.2.jar 包。

4.2　第一个 Spring 程序

项目案例：之前在一个类中调用另一个类都是在本类中先新建一个要调用的另一个类对象，再调用其方法，这次用 Spring 实现在一个类中不新建另一个类的对象也能调用到另一个类的程序。（项目源码参见本书配套资源：第 4 章/第一个 spring 程序/spring1）

思路分析：将用到 Spring 的控制反转，一个类无须自己创建另一个类的对象（实例)，而是通过 Spring 容器来构建另一个类的实例。

实现步骤：

（1）创建 Dynamic Web Project 项目 spring1，导入 JAR 包，所需的 JAR 包如图 4.3 所示。

第 4 章　Spring 入门

图 4.3　导入 JAR 包

其中以 Spring 开头的 JAR 包都可在 4.1 节下载的解压缩文件的 libs 中找到，其他的可自行上网下载，将这些 JAR 包逐一复制到目录 WebContent/WEB-INF/lib 下，然后选中这些 JAR 包，单击鼠标右键，在弹出的菜单中选择"build path"→"add to build path"。

（2）创建包 com.lifeng，在该包下创建类 Login，代码如下：

```
public class Login {
    private String name;
    public String getName() {
        return name;
    }
    public void setName(String name) {
        this.name = name;
    }
    public void show(){
        System.out.println(name+"你好!欢迎你登录!");
    }
}
```

【注意】属性 private String name 要有 setter 方法。

（3）创建 Spring 的配置文件 apllicationContext.xml。

在 src 下新建一个 XML 文件，命名为 apllicationContext.xml，建完后，只有空的 XML 结构，如下所示：

`<?xml version="1.0" encoding="UTF-8"?>`

接下来需要在配置文件的头部添加一些约束信息，先在之前下载的压缩包解压后的 docs 目录下找到 spring-framework-reference/html/xsd-configuration.html 文件，双击打开文件，找到图 4.4 所示的内容。

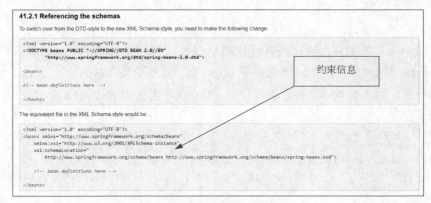

图 4.4　找到配置文件约束

图 4.4 中下部分方框内的 `<beans></beans>` 标签内的代码即要复制的约束信息，把它复制粘贴到配

79

置文件中，配置文件的代码如下所示：
```xml
<?xml version="1.0" encoding="UTF-8"?>
<beans xmlns="http://www.springframework.org/schema/beans"
    xmlns:xsi="http://www.w3.org/2001/XMLSchema-instance"
    xsi:schemaLocation="
       http://www.springframework.org/schema/beans
http://www.springframework.org/schema/beans/spring-beans.xsd">
    <!-- bean definitions here -->
</beans>
```

接下来在配置文件时实现创建实例化的对象的功能，在<beans></beans>之间，即上面代码的<!-- bean definitions here -->位置（删除它）添加一对<bean/>节点，最终内容如下：

```xml
<?xml version="1.0" encoding="UTF-8"?>
<beans xmlns="http://www.springframework.org/schema/beans"
    xmlns:xsi="http://www.w3.org/2001/XMLSchema-instance"
    xsi:schemaLocation="
       http://www.springframework.org/schema/beans
http://www.springframework.org/schema/beans/spring-beans.xsd">
    <!-- 相当于传统的 Login login =new Login() -->
    <bean id="login" class="com.lifeng.Login"/>
</beans>
```

（4）创建测试类 Test，关键代码如下：

```java
public static void main(String[] args) {
    //读取配置文件
    ApplicationContextcontext=new ClassPathXmlApplicationContext("applicationContext.xml");
    //从配置文件中获取实例Bean
    Login login=(Login) context.getBean("login");
    //使用Bean
    login.show();
}
```

其中，语句 Login login=(Login)context.getBean("login")用于获取配置文件中创建好的 Login 类的实例化对象 login。也就是说测试类 Test 无须再自己创建 Login 类的对象，因为配置文件已经帮它创建好了，只要拿过来用就行。对比下不使用 Spring 时的测试类代码：

```java
public static void main(String[] args) {
    Login login=new Login();
    login.show();
}
```

虽然不用 Spring 框架的这段看起来似乎更简洁，但它们有一个本质的区别，就是测试类 Test 需要自行创建 Login 类的对象，使 Test 类依赖于 Login 类，使用了 Spring 的测试类 Test 则无须自行创建 Login 类的对象，而是 Spring 已帮它创建好了，Test 类不再直接依赖于 Login 类。

第一次测试结果：

```
log4j:WARNNoappenderscouldbefoundforlogger (org.springframework.core.env.StandardEnvironment).
log4j:WARN Please initialize the log4j system properly.
log4j:WARN See http://logging.apache.org/log4j/1.2/faq.html#noconfig for more info.
Null 你好！欢迎你登录！
```

结果有了输出，证明 Spring 框架总体运行成功，但还有一个小问题，就是输出的 name 为 null，这是因为 name 属性没有赋值，默认值是 null。那么怎么给 name 属性赋值呢？传统赋值方法是先新建一个对象，再调用其 setter 方法。假如使用传统方法，则 Test 测试类中可用如下代码实现对象的实例化和赋值：

```
Login  login =new Login()
Login.setName("张三");
```

另一个方法就是在 Spring 配置文件中完成对属性的赋值，即创建 Bean 对象的同时实现属性的赋值。

（5）对 Spring 的配置文件做如下修改：

```xml
<?xml version="1.0" encoding="UTF-8"?>
<beans xmlns="http://www.springframework.org/schema/beans"
    xmlns:xsi="http://www.w3.org/2001/XMLSchema-instance"
    xsi:schemaLocation="
    http://www.springframework.org/schema/beans
    http://www.springframework.org/schema/beans/spring-beans.xsd">
<!-- 相当于在程序中创建一个 Login 类的对象,对象名为 login,同时给其中的属性 name 赋值张三 -->
    <bean id="login" class="com.lifeng.Login">
     <property name="name">
            <value>张三</value>
        </property>
    </bean>
</beans>
```

第二次运行测试 Test，结果如下：

张三你好！欢迎你登录！

可见，name 已经成功输出，证明在 Spring 中设置属性值成功。

总结：通过这个案例，可以发现测试类 Test 不再直接创建（new）另外一个类 Login 的实例化对象，而是改由 Spring 来创建 Login 类的实例化对象，甚至该实例化对象的属性值都由 Spring 进行赋值，这样 Test 类就不再直接依赖于 Login 类，可谓实现了 Test 类与 Login 类的"解耦"。Spring 成为类与类之间实现"解耦"的关键第三方。

上机练习

小鸟类有 color 属性、fly()方法，其中 fly()方法用于输出"XX 颜色的小鸟在天上飞"。测试类 Test，使用 Spring 实现在 Test 中调用小鸟类的 fly()方法的程序。

思考题

1. Spring 有哪几个模块？
2. Spring 实现"解耦"的方式有哪些？

第 5 章　Spring 控制反转

本章目标

- 理解 IoC 的含义
- 理解依赖注入的含义
- 掌握实现依赖注入的方法
- 掌握自动装配的相关知识
- 学会使用注解注入

IoC（Inversion of Control，控制反转），即调用者创建被调用者的实例对象不是由调用者自己完成，而是由 Spring 容器完成。

使用 IoC 后，一个对象（调用者）依赖的其他对象（被调用者）会通过被动的方式传递进来，而不是这个对象（调用者）自己创建或者查找依赖对象（被调用者）。Spring 容器在对象初始化时不等对象请求就主动将依赖对象传递给它。IoC 只是一种编程思想，具体的实现方法是依赖注入。

依赖注入（Dependency Injection，DI）是指程序在运行过程中，若需要调用另一个对象协助，无须在代码中用硬编码（如用 new 构造方法）创建被调用者，而是依赖于外部容器（如 Spring），被调用者由外部容器创建后传递给程序。依赖注入解决了传统的编程方法中类与类之间严重的直接依赖的问题，是目前最优秀的解耦方式。依赖注入让 Spring 的 Bean 之间以配置文件的方式组织在一起，而不是以硬编码的方式耦合在一起的。

5.1 依赖注入

下面演示一个传统的存在直接依赖的程序，然后分析它的问题，再改用 Spring 来解决问题。

项目案例：一个传统的存在直接依赖的分层构架项目。（项目源码参见本书配套资源：第 5 章/依赖注入/spring2）

实现步骤：

（1）创建项目 spring2，导入 Spring 所需的 JAR 包，并创建 RegisterDao 接口和 RegisterDaoImpl 实现类，具体如下：

```
package com.lifeng.dao;
public interface RegisterDao {
    public void regist();
}
public class RegisterDaoImpl implements RegisterDao{
    @Override
    public void regist() {
        System.out.println("注册成功,数据已存入数据库!");
    }
}
```

（2）创建 RegisterService 类，代码如下：

```
public class RegisterService {
    RegisterDao registerDao;
    public void regist(){
        registerDao =new RegisterDaoImpl();
        registerDao.regist();
        //省略getter,setter方法
    }
}
```

使用传统方法实例化另一个类，存在直接依赖

（3）创建测试类 TestRegister，代码如下：

```
public class TestRegister {
    public static void main(String[] args) {
        RegisterService regService=new RegisterService();
        regService.regist();
    }
}
```

问题分析：RegisterService 类在 regist() 方法中创建了另一个类 RegisterDaoImpl 的实例化对象，并调用其 regist() 方法，这种情况属于 RegisterService 类直接依赖 RegisterDaoImpl 类，假如 RegisterDaoImpl 类有变化，将影响 RegisterService 类。举个简单的例子，假如类 RegisterDaoImpl 的名称改成了 RegisterDaoImpl888，则 RegisterService 类中的 regist() 方法也会随之修改。

```
public void regist(){
    registerDao=new RegisterDaoImpl888();   ← new 后面的类名跟着改变
    registerDao.regist();
}
```

如果受依赖的类和方法还有更多，则都要修改，这就是依赖性造成的一个不利影响。Spring 框架利用依赖注入可以很好地解决这个问题。下面来看用 Spring 改造上述项目的具体步骤。

（1）在项目 spring2 中导入 Spring JAR 包，具体的 JAR 包请参见第 4 章。

（2）创建 Spring 的配置文件 apllicationContext.xml，添加约束，实现创建实例化对象的功能，代码如下：

```xml
<?xml version="1.0" encoding="UTF-8"?>
<beans xmlns="http://www.springframework.org/schema/beans"
    xmlns:xsi="http://www.w3.org/2001/XMLSchema-instance"
    xsi:schemaLocation="
        http://www.springframework.org/schema/beans
http://www.springframework.org/schema/beans/spring-beans.xsd">
    <!-- 相当于在程序中创建一个 RegisterDaoImpl 类的实例化对象，对象名为 registerDao -->
    <bean id="registerDao" class="com.lifeng.dao.RegisterDaoImpl"/>
    <!-- 相当于在程序中创建一个 RegisterService 类的对象，对象名为 regService -->
    <bean id="regService" class="com.lifeng.service.RegisterService">
        <!--为对象 regService 中的 registerDao 属性注入实例化对象 registerDao -->
        <property name="registerDao" ref="registerDao"/>
</bean>
</beans>
```

（3）修改 RegisterService 类，注释或删除 regist() 方法中创建 RegisterDaompl 实例的代码如下：

```java
public class RegisterService {
    private RegisterDao registerDao; //省略getter,setter
    public void regist(){
        //registerDao=new RegisterDaoImpl();
        registerDao.regist();
    }
}
```

【注意】需要为 registerDao 属性创建 setter 方法。

在这一步骤中，语句 registerDao =new RegisterDaoImpl();被注释掉了，也就是说这个 registerDao 对象表面上是没有被实例化的，按传统方法进行下去，最后一定会报错，但这里不用担心，因为在步骤（2）的 Spring 配置文件中已经明确地给 registerDao 注入了实例化对象。

（4）修改测试类，具体如下：

```java
public static void main(String[] args) {
    ApplicationContextcontext=new ClassPathXmlApplicationContext("applicationContext.xml");
    RegisterService regService=(RegisterService) context.getBean("regService");
    regService.regist();
}
```

运行测试结果如下：

注册成功，数据已存入数据库！

类 RegisterService 不再直接依赖类 RegisterDaoImpl，类 RegisterService 里面虽然要调用 RegisterDaoImpl 的实例，但这个实例无须 RegisterService 自己创建，而是配置文件中已经用代码<bean id="registerDao"class="com.lifeng.dao.RegisterDaoImpl">创建好了，然后又创建了一个类 RegisterService 的实例化对象 regService，并为对象 regService 中的 registerDao 属性注入实例化对象 registerDao。

接下来看如果改变 RegisterDaoImpl 的名字，类 RegisterService 需不需要做相应的改变。

（1）先修改类 RegisterDaoImpl 的名字为 RegisterDaoImpl888。

（2）再修改配置文件如下：

```
<?xml version="1.0" encoding="UTF-8"?>
<beans xmlns="http://www.springframework.org/schema/beans"
    xmlns:xsi="http://www.w3.org/2001/XMLSchema-instance"
    xsi:schemaLocation="http://www.springframework.org/schema/beans
http://www.springframework.org/schema/beans/spring-beans.xsd">
    <!-- 相当于在程序中创建一个 RegisterDaoImpl 类的对象，对象名为 registerDao -->
    <bean id="registerDao" class="com.lifeng.dao.RegisterDaoImpl888"/>
    <!-- 相当于在程序中创建一个 RegisterService 类的对象，对象名为 regService -->
    <bean id="regService" class="com.lifeng.service.RegisterService">
    <!-- 相当于为对象 regService 中的 registerDao 属性注入实例化对象 registerDao -->
     <property name="registerDao" ref="registerDao"/>
    </bean>
</beans>
```

文件只改了<bean id="registerDao" class="com.lifeng.dao.RegisterDaoImpl888"/>这一处，其他地方没变，类 RegisterService 和测试类均不做任何改动。

（3）测试运行，发现结果一样。证明了类 RegisterService 不再直接依赖类 RegisterDaoImpl。

5.2　Spring 配置文件中 Bean 的属性

Spring 配置文件的根节点<beans>下一级节点是<bean>，用于创建 Java 类的实例化对象（俗称 Bean）。<bean>有多个属性及子节点，如表 5.1 所示。

表 5.1　<bean>的属性

<bean>的属性与子节点	说明
id	唯一标识一个 Bean，相当于创建的实例化对象的名称
class	全限定性类名，该类用于创建实例化对象
scope	Bean 的作用范围，常见的取值有 singleton（单例）、prototype（原型）、request、session、global Session、application、websocket，默认值是 singleton，表示程序中每次调用 getBean()方法获取的 Bean 都是同一个对象
property	是<bean>节点的子节点，用于调用 setter 方法给 Bean 的各个属性赋值
contructor-arg	是<bean>节点的子节点，用于传递构造方法的参数，再调用构造方法实现 Bean 的实例化
value	既可作为上面两种子节点的属性，也可作为它们的子节点，用于直接指定一个常量值
ref	property、contructor-arg 这两种子节点的属性或子节点，用于引用另一个 Bean
set	用于 Set 集合类型的属性的赋值
list	用于 List 集合类型的属性的赋值
map	用于 Map 集合类型的属性的赋值

5.3　Bean 的作用域

当通过 Spring 容器创建一个 Bean 实例时，不仅可以完成 Bean 的实例化，还可以通过 scope 属性为 Bean 指定特定的作用域。Spring 支持 5 种作用域。

- singleton：单例模式。在整个 Spring 容器中，使用 singleton 定义的 Bean 将是单例的，即只有一个实例，第二次以后调用的 Bean 与第一次调用的是同一个对象，默认为单例的。该 Bean 是在容器被创建时即被装配好了。
- prototype：原型模式，即每次使用 getBean()方法获取的同一个 Bean 的实例都是一个新的实例。对于 scope 为 prototype 的原型模式，在代码中使用该 Bean 实例时才对其进行装配。
- request：对于每次 HTTP 请求，都将会产生一个不同的 Bean 实例。
- session：对于每个不同的 HTTP session，都将产生一个不同的 Bean 实例。
- globalSession：所有的 HTTP Session 共享同一个 Bean 实例。

其中，singleton 和 prototype 最为常用，下面通过一个案例介绍这两者的区别。

项目案例：创建 RegisterService 类的两个 Bean，第一个 Bean 的 id 是 regService1，scope 属性值是 singleton，第二个 Bean 的 id 是 regService2，scope 属性值是 prototype。测试类中各调用 getBean()方法两次，观察每次获取的对象是不是同一个对象。（项目源码参见本书配套资源：第 5 章/Bean 的作用域/spring3）

实现步骤：

（1）将项目 spring2 复制为 spring3，在 Spring 配置文件中添加两个 Bean，其中一个作用域是 singleton，另一个是 prototype。具体代码如下：

```xml
<?xml version="1.0" encoding="UTF-8"?>
<beans xmlns="http://www.springframework.org/schema/beans"
    xmlns:xsi="http://www.w3.org/2001/XMLSchema-instance"
    xsi:schemaLocation="
        http://www.springframework.org/schema/beans
http://www.springframework.org/schema/beans/spring-beans.xsd">
    <!-- 相当于在程序中创建一个 RegisterDaoImpl 类的对象，对象名为 registerDao -->
    <bean id="registerDao" class="com.lifeng.dao.RegisterDaoImpl"/>
    <!-- 相当于在程序中创建一个 RegisterService 类的对象，对象名为 regService1 -->
    <bean id="regService1" class="com.lifeng.service.RegisterService"
scope="singleton">
    <!-- 相当于为对象 regService1 中的 registerDao 属性注入实例化对象 registerDao -->
    <property name="registerDao" ref="registerDao"/>
    </bean>
    <bean id="regService2" class="com.lifeng.service.RegisterService"
scope="prototype">
    <!-- 相当于为对象 regService2 中的 registerDao 属性注入实例化对象 registerDao -->
    <property name="registerDao" ref="registerDao"/>
    </bean>
</beans>
```

（2）创建测试类 TestRegister，关键代码如下：

```java
public static void main(String[] args) {
    ApplicationContext context=new ClassPathXmlApplicationContext("applicationContext.xml");
    RegisterService regService11=(RegisterService) context.getBean("regService1");
        RegisterService regService12=(RegisterService) context.getBean("regService1");
        System.out.println("regService11 与 regService12 是否同一个对象:"+(regService11
```

```
==regService12));
            RegisterServiceregService21=(RegisterService) context.getBean("regService2");
            RegisterServiceregService22=(RegisterService) context.getBean("regService2");
            System.out.println("regService21 与 regService22 是否同一个对象:"+(regService21
==regService22));
    }
```

（3）测试结果如下：

regService11 与 regService12 是否同一个对象:true
regService21 与 regService22 是否同一个对象:false

测试结果证明：当 scope 属性的值为 singleton 时，每次 getBean()获取到的对象都是同一个，而当 scope 属性的值为 prototype 时，每次 getBean()方法获取的对象都是一个新的对象。

5.4 基于 XML 的依赖注入

Bean 的依赖注入又称为 Bean 的装配，即利用 Spring 实现 Bean 的实例化过程，在这个过程中需完成实例化对象的创建，并给对象的各个属性赋值等。Spring 有多种 Bean 的装配方式，包括基于 XML 文件的装配、基于注解的装配和自动装配。

其中，基于 XML 文件的配置又分为设值注入和构造注入两种装配方式。

Spring 实例化 Bean 的时候，首先会调用 Bean 默认的无参构造方法来实例化一个空值的 Bean 对象，接着对 Bean 对象的属性进行初始化。初始化是由 Spring 容器自动完成的，称为注入。有下列多种注入方式。

5.4.1 设值注入

设值注入是指 Spring 通过反射机制调用 setter 方法来注入属性值。Bean 必须满足以下两点要求才能被实例化。

① Bean 类必须提供一个无参的构造方法。注意，如果程序员定义了有参的构造方法，则必须要显式地提供无参的构造方法。

② 属性需要提供 setter 方法。

项目案例：通过设值注入，实现 Bean 对象的实例化。（项目源码参见本书配套资源：第 5 章/基于 xml 的依赖注入/spring4）

实现步骤：

（1）复制 spring3 为 spring4，删除 com.lifeng.dao 和 com.lifeng.service 包，删除测试类。创建包 com.lifeng.entity，并在该包下创建实体类 User 代码如下：

```
public class User {
    private String uid;
    private String uname;
    private String gender;
    private int age;
    //需要显式地给出无参构造方法
    public User(){
    }
```

```java
        public User(String uname,String gender,int age){
            this.uname=uname;
            this.gender=gender;
            this.age=age;
        }
        public User(String uid,String uname,String gender,int age){
            this.uid=uid;
            this.uname=uname;
            this.gender=gender;
            this.age=age;
        }
        public void show(){
            System.out.println("用户编号:"+uid+" 用户姓名:"+uname+" 性别:"+gender+" 年龄:"+age);
        }
        //省略 setter、getter 的方法
}
```

（2）修改配置文件，删除原有的 Bean，按以下格式新建 Bean：

```xml
<?xml version="1.0" encoding="UTF-8"?>
<beans xmlns="http://www.springframework.org/schema/beans"
    xmlns:xsi="http://www.w3.org/2001/XMLSchema-instance"
    xsi:schemaLocation="
        http://www.springframework.org/schema/beans
http://www.springframework.org/schema/beans/spring-beans.xsd">
    <!-- 设值注入第一种格式 -->
    <bean id="user1" class="com.lifeng.entity.User">
        <property name="uid" value="1"/>
        <property name="uname" value="张三"/>
        <property name="gender" value="男"/>
        <property name="age" value="18"/>
    </bean>
</beans>
```

通过子节点<property>对 User 对象里面的所有属性赋值，可以用<property>的属性 value 来赋常量值，也可以改为通过<property>的下一级子节点<value>来赋值，下面用后一种方式再创建一个 Bean：

```xml
<!-- 设值注入第二种格式 -->
<bean id="user2" class="com.lifeng.entity.User">
    <property name="uid">
        <value>2</value>
    </property>
    <property name="uname">
        <value>李四</value>
    </property>
    <property name="gender">
        <value>女</value>
    </property>
    <property name="age">
        <value>19</value>
    </property>
</bean>
```

两者效果相同。

（3）删除原有测试类，新建测试类 TestUser1，关键代码如下：
```
public static void main(String[] args) {
        ApplicationContextcontext=new ClassPathXmlApplicationContext("applicationContext.xml");
        User user1=(User) context.getBean("user1");
        user1.show();
        User user2=(User) context.getBean("user2");
        user2.show();
    }
```
测试结果如下：

用户编号:1 用户姓名:张三性别:男年龄:18
用户编号:2 用户姓名:李四性别:女年龄:19

5.4.2 构造注入

构造注入是指在构造调用者实例的同时，完成被调用者的实例化，即使用构造方法进行赋值。

（1）在项目 spring4 中修改配置文件，继续创建一个 id 为 user3 的 Bean。
```
<bean id="user3" class="com.lifeng.entity.User">
    <constructor-arg name="uid" value="3"/>
    <constructor-arg name="uname" value="张无忌"/>
    <constructor-arg name="gender" value="男"/>
    <constructor-arg name="age" value="22"/>
</bean>
```
<constructor-arg/>标签中用于指定参数的属性如下。

① name：指定构造方法中的参数名称。

② index：指明该参数对应构造器的第几个参数，从 0 开始，该属性不要也可，但注意赋值顺序要与构造器中的参数顺序一致。

如果使用 index 属性，则用以下代码进行注入。
```
<bean id="user4" class="com.lifeng.entity.User">
    <constructor-arg index="0" value="4"/>
    <constructor-arg index="1" value="张无忌"/>
    <constructor-arg index="2" value="男"/>
    <constructor-arg index="3" value="22"/>
</bean>
```
此外，<constructor-arg/>的 type 属性可用于指定其类型。基本类型直接写类型关键字，非基本类型需要写全限定性类名。

（2）创建测试类 TestUser2。
```
public class TestUser2 {
    public static void main(String[] args) {
        ApplicationContextcontext=new ClassPathXmlApplicationContext("applicationContext.xml");
        User user3=(User) context.getBean("user3");
        user3.show();
        User user4=(User) context.getBean("user4");
        user4.show();
    }
}
```

测试结果，两种格式效果相同。
用户编号:3 用户姓名:张无忌性别:男年龄:22
用户编号:4 用户姓名:张无忌性别:男年龄:22

5.4.3　p 命名空间注入

p 命名空间可以简化属性值的注入。

（1）首先在项目 spring4 的 Spring 配置文件头中引入 p 命名空间：

```
<?xml version="1.0" encoding="UTF-8"?>
<beans xmlns="http://www.springframework.org/schema/beans"
    xmlns:xsi="http://www.w3.org/2001/XMLSchema-instance"
    xmlns:p="http://www.springframework.org/schema/p"          ← 引入 p 命名空间
    xsi:schemaLocation="http://www.springframework.org/schema/beans
http://www.springframework.org/schema/beans/spring-beans.xsd">
```

上面第 4 行代码即为添加的 p 命名空间的声明。

（2）在 Spring 配置文件中再添加一个 Bean，使用 p 命名空间注入属性值：

```
<bean id="user5" class="com.lifeng.entity.User"
p:uid="5" p:uname="王五" p:gender="女" p:age="23"/>
```

p 命名空间注入也是采用设值注入方式，故需要有相应属性的 setter 方法。通过上面代码可知，对照普通的设值注入，p 命名空间注入简化了很多，其基本语法参见表 5.2。

表 5.2　p 命名空间注入语法

语法格式	说明
p:bean 属性="值"	该值为基本类型的值
p:bean 属性-ref="值"	该值为其他 Bean 的 id，ref 称为引用，即引用了其他 Bean

（3）创建测试类 TestUser3，具体如下：

```
public class TestUser3 {
    public static void main(String[] args) {
        ApplicationContextcontext=new ClassPathXmlApplicationContext("applicationContext.xml");
        User user5=(User) context.getBean("user5");
        user5.show();
    }
}
```

测试结果如下：
用户编号:4 用户姓名:王五 性别:女 年龄:23

5.4.4　各种数据类型的注入

1. 注入常量

使用<value></value>子节点，将常量（字符串及基本数据类型）置于中间，如 5.4.1 节的设值注入的第二种格式所示。也可以不用<value>子节点，而是直接在<property>节点中嵌入 value 属性，如 5.4.1 节的设值注入的第一种格式所示。

注入字符串常量值如果用到特殊字符，则需要替换为表 5.3 对应的表达式。

表 5.3 特殊字符

符号	表达式
>	&gt;
<	&lt;
&	&amp;
'	&apos;
"	&quot;

2. 引用其他 Bean 组件

在 Spring 中定义的各个 Bean 之间可以互相引用（注入），用于建立依赖关系，可以使用 ref 属性实现，也可使用<ref>子节点实现。

项目案例：新建一个 Company 类，定义公司的编号、名称、地址属性，并修改用户 User 类，添加一个 Company 类型的属性，用于设置用户所在的公司。（项目源码参见本书配套资源：第 5 章/基于 xml 的依赖注入/spring4）

实现步骤：

（1）在项目 spring4 中创建 Company 类，代码如下：

```
public class Company {
    private int id;
    private String companyname;
    private String address;
    //省略 getter、setter 方法
    public String toString(){
        return "公司名称:"+companyname+" ,公司地址:"+address;
    }
}
```

（2）修改 User 类，添加 Company 属性及 getter 和 setter 方法，代码如下：

```
private Company company;
public Company getCompany() {
    return company;
}
public void setCompany(Company company) {
    this.company = company;
}
```

（3）在配置文件中创建 Company 类的 Bean，代码如下：

```
<bean id="mycompany" class="com.lifeng.entity.Company">
    <property name="id">
        <value>1</value>
    </property>
    <property name="companyname">
        <value>砺锋科技公司</value>
    </property>
    <property name="address">
        <value>广州天河软件园</value>
    </property>
</bean>
```

（4）创建 User 类的 Bean，引用 Company 类的 Bean，代码如下：

```xml
<bean id="user6" class="com.lifeng.entity.User">
    <property name="uid">
        <value>6</value>
    </property>
    <property name="uname">
        <value>李四</value>
    </property>
    <property name="gender">
        <value>女</value>
    </property>
    <property name="age">
        <value>19</value>
    </property>
    <property name="company">
        <ref bean="mycompany"/>    ← 引用（注入）了另外一个 Bean
    </property>
</bean>
```

可用<ref>标签引用其他的 Bean，也可以直接在<property>中用 ref 属性实现同样的效果：

```xml
<property name="company" ref ="mycompany"/>
```

（5）测试类 TestUser4，关键代码如下：

```java
public static void main(String[] args) {
    ApplicationContext context=new ClassPathXmlApplicationContext("applicationContext.xml");
    User user6=(User) context.getBean("user6");
    user6.show();
    System.out.println("所在公司:\n"+user6.getCompany());
}
```

测试结果：

用户编号:6 用户姓名:李四 性别:女 年龄:19
所在公司:
公司名称:砺锋科技公司 ,公司地址:广州天河软件园

5.5 自动注入

对域属性（如上述项目中 User 类中的 company 属性就是一个域属性）的注入，也可以不在配置文件中显式注入。可以通过为<bean>标签设置 autowire 属性值，从而为域属性进行隐式自动注入（又称自动装配）。根据自动注入判断标准的不同，自动注入可以分为两种方式。

（1）byName：根据名称自动注入。

（2）byType：根据类型自动注入。

5.5.1 byName 方式自动注入

当配置文件中被调用者的 Bean 的 id 值与调用者的 Bean 类的属性名相同时，可使用 byName 方式，让容器自动将被调用者的 Bean 注入调用者的 Bean。容器通过调用者的 Bean 类的属性名与配置文件的被调用者的 Bean 的 id 进行比较而实现自动注入。

项目案例：先创建一个 Company 类的 Bean，其 id 为 company，再创建一个 User 类型的 Bean，其 id 为 user7。user7 这个 Bean 中的 company 属性的注入不再直接引用 company，而是改为自动注入。（项目源码参见本书配套资源：第 5 章/基于 xml 的依赖注入/spring4）

实现步骤：

（1）在项目 spring4 的 Spring 配置文件中创建 Company 类的 Bean，其 id 为 company，代码如下：

```xml
<bean id="company" class="com.lifeng.entity.Company">
    <property name="id">
        <value>2</value>
    </property>
    <property name="companyname">
        <value>砺锋软件公司</value>
    </property>
    <property name="address">
        <value>广州天河建中路</value>
    </property>
</bean>
```

（2）创建一个 User 类型的 Bean，采用自动注入，代码如下：

```xml
<bean id="user7" class="com.lifeng.entity.User" autowire="byName">
    <property name="uid">
        <value>7</value>
    </property>
    <property name="uname">
        <value>李四</value>
    </property>
    <property name="gender">
        <value>女</value>
    </property>
    <property name="age">
        <value>19</value>
    </property>
</bean>
```

（autowire="byName" → 自动注入（按 Bean 名称））

（这里无须再给 company 属性赋值）

这个 Bean 并没有出现直接为 User 类的 company 属性注入值的有关配置，但值得注意的是在这个 Bean 的头部出现了一个 autowire 属性，就是自动装配的意思。这时 Spring 会自动检查该 Bean 下的域属性，并自动给它们赋值。这里是按 Bean 的名称赋值，如果这个域属性名叫 company，则 Spring 会在配置文件中自动查找同样名为 company 的 Bean，把名为 company 的 Bean 注入这个名为 company 的域属性。如果当前项目并不存在名为 company 的 Bean，则什么也不干，也不会报错。如果不采用自动装配，则需要显式地给所有域属性注入值。

（3）测试类 TestUser5 的代码如下：

```java
public class TestUser5 {
    public static void main(String[] args) {
        ApplicationContext context=new ClassPathXmlApplicationContext("applicationContext.xml");
        User user7=(User) context.getBean("user7");
        user7.show();
        System.out.println("所在公司：\n"+user7.getCompany());
    }
}
```

测试结果如下：

用户编号:7 用户姓名:李四 性别:女 年龄:19

所在公司：

公司名称:砺锋软件公司 ,公司地址:广州天河建中路

5.5.2　byType 方式自动注入

byType 注入是指按类型进行匹配注入，即只要 Spring 容器中有跟域属性类型相同的 Bean 就自动注入。使用 byType 方式自动注入，要满足条件：配置文件中的被调用者 Bean 的 class 属性指定的类，要与调用者的 Bean 类的某域属性的类型同源，要么相同，要么有 is-a 关系（子类或是实现类）。但这样同源的被调用的 Bean 只能有一个。多于一个，容器就不知该匹配哪一个，从而导致报错。（项目源码参见本书配套资源：第 5 章/基于 xml 的依赖注入/spring4）

关键步骤：

（1）在项目 spring4 中新建一个 Bean，id 为 user8，把装配方式改为 byType，其他基本与 user7 一样，具体如下：

```
<bean id="user8" class="com.lifeng.entity.User" autowire="byType">
  <-- 省略其他 -->
</bean>
```

（2）创建测试类 TestStudent6，代码如下：

```
public static void main(String[] args) {
        ApplicationContextcontext=new ClassPathXmlApplicationContext("applicationContext.xml");
        User user8=(User) context.getBean("user8");
        user8.show();
        System.out.println("所在公司:\n"+user8.getCompany());
    }
```

此时运行测试程序会报错：

```
Exceptioninthread"main" org.springframework.beans.factory.UnsatisfiedDependencyException: Error creating bean with name 'user8' defined in class path resource [applicationContext.xml]: Unsatisfied dependency expressedthroughbeanproperty'company'; nestedexceptionis org.springframework.beans.factory.NoUniqueBeanDefinitionException: No qualifying bean of type 'com.lifeng.entity.Company' available: expected single matching bean but found 2: mycompany,company
```

这是因为这个项目中 class="com.lifeng.entity.Company"的 Bean 目前有两个（一个砺锋科技 mycompany，一个砺锋软件 company，都是同源），Spring 不知该装配哪个。删掉其中一个，则程序正常运行。这里注释掉 id 为 company 的 Bean，运行结果如下：

用户编号:8 用户姓名:李四 性别:女 年龄:19

所在公司：

公司名称:砺锋科技公司 ,公司地址:广州天河软件园

5.6　Spring 配置文件的拆分

在项目中，随着应用规模的增加，Bean 的数量也不断增加，配置文件会变得越来越庞大、臃肿，可读性变差。为了提高配置文件的可读性与可维护性，可以将 Spring 配置文件分解成多个配置文件。

拆分策略有两种，一是拆分为平等关系的若干个配置文件，二是拆分为包含关系的若干个配置文件。

5.6.1 拆分为若干个平等关系的配置文件

将配置文件分解为地位平等的多个配置文件，各配置文件之间为不分主次的并列关系。在代码中将所有配置文件的路径定义为一个 String 数组，将其作为容器初始化参数出现。

例如，在 src 下创建 3 个独立的 Spring 配置文件，名称分别为：spring-base.xml、spring-dao.xml、spring-service.xml，分别负责基本信息、数据库连接信息、业务逻辑层信息的配置。应用程序的运行需要同时用到这 3 个配置文件。用下面代码一次性调用 3 个配置文件（项目源码中的代码仅供参考，未实际配置）：

```
public class TestStudent7 {
    public static void main(String[] args) {
        String[]resources={"spring-base.xml","spring-dao.xml","spring-service.xml"};
        ApplicationContext context=new ClassPathXmlApplicationContext(resources);
        User user8=(User) context.getBean("user8");
        user8.show();
        System.out.println("所在公司:\n"+user8.getCompany());
    }
}
```

> 使用数组列出 3 个配置文件的名称，这样就可以一次性读出 3 个配置文件

5.6.2 拆分为父子关系的若干个配置文件

各配置文件中有一个总配置文件（父级），总配置文件将其他子文件（子级）通过<import>引入。在代码中只需要使用总配置文件对容器进行初始化即可。

例如，在 src 下创建 4 个 Spring 配置文件，名称分别为：spring-base.xml、spring-dao.xml、spring-service.xml、spring-all.xml。

其中，spring-all.xml 是总配置文件，在它里面通过下列代码可引入其他 3 个配置文件：

```
<import rsource="classpath: spring-base.xml"></import>
<import rsource="classpath: spring-dao.xml"></import>
<import rsource="classpath: spring-service.xml"></import>
```

也可使用通配符 "*"，即上面三行可以利用通配符综合成一行：

```
<import rsource="classpath: spring-*.xml"></import>
```

【注意】这种情况下要求总配置文件名本身不能满足 "*" 所能匹配的格式，否则将出现循环递归包含。就本例而言，父配置文件不能匹配 spring-*.xml 的格式，即不能起名为 spring-all.xml。

5.7 基于注解的依赖注入

除了用 XML 配置方式进行依赖注入外，还可以使用注解直接在类中定义 Bean 实例，这样就不再需要在 Spring 配置文件中声明 Bean 实例。使用注解，除了原有 Spring 配置，还要注意以下 3 个关键步骤。

项目案例：使用注解定义和使用 Bean。（项目源码参见本书配套资源：第 5 章/基于注解的依赖注入/spring5）

实现步骤：

（1）复制 spring4 为 spring5，删除所有测试类，删除配置文件内容。导入 AOP 的 JAR 包，包名为 spring-aop-4.3.4.RELEASE.jar，这是因为注解的后台实现用到了 AOP 编程。

（2）需要更换配置文件头，即添加相应的约束。约束在 Spring 的解压文件夹的\docs\spring-framework-reference\html\xsd-configuration.html 文件中，如图 5.1 所示。将相关代码复制到配置文件中。

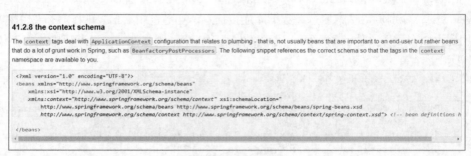

图 5.1 使用注解时的配置文件约束

（3）需要在 Spring 配置文件中配置组件扫描器，用于在指定的基本包中扫描注解，代码如下：

```
<?xml version="1.0" encoding="UTF-8"?>
<beans xmlns="http://www.springframework.org/schema/beans"
    xmlns:xsi="http://www.w3.org/2001/XMLSchema-instance"
    xmlns:context="http://www.springframework.org/schema/context" xsi:schemaLocation="
    http://www.springframework.org/schema/beans http://www.springframework.org/schema/beans/spring-beans.xsd
    http://www.springframework.org/schema/context http://www.springframework.org/schema/context/spring-context.xsd"> <!-- bean definitions here -->

    <context:component-scan base-package="com.lifeng"/>

</beans>
```

（配置组件扫描器，在这个基本包中扫描注解）

5.7.1 使用注解@Component 定义 Bean

注解@Component 的 value 属性用于指定该 Bean 的 id 值。例如在实体类 User 上添加注解@Component(value="user")，它的意思是创建一个 User 类的 Bean 实例，Bean 的 id 为 user。注意其放置的具体位置。

在项目 spring5 的 User 类中添加@Component 注解，代码如下：
```
@Component("user")
public class User {
    private String uid;
    private String uname;
    private String gender;
    private int age;
    private Company company;
    //省略其他方法
}
```

【注意】 @Component("user")等同于@Component(value="user")，即默认的属性是 value，其效果等同于 xml 配置文件：

```
<bean id="user" class="com.lifeng.entity.User" />
```

Spring 还另外提供了 3 个与@Component 具有相同功能的注解。
- @Repository：专用于对 DAO 层实现类进行注解。
- @Service：专用于对 Service 层实现类进行注解。
- @Controller：专用于对 Controller 类进行注解。

5.7.2 Bean 的作用域@Scope

需要在类上使用注解@Scope，其 value 属性用于指定作用域，默认值为 singleton。
在项目 spring5 的 Student 类中添加下列注解：

```
@Scope("prototype")
@Component("student")
public class User {
//省略其他代码
}
```

在上述代码中，@Scope("prototype")等同于@Scope(value="prototype")，这样就设置了 Bean 的作用范围为"prototype"。

5.7.3 基本类型属性注入@Value

需要在属性上使用注解@Value，该注解的 value 属性用于指定要注入的值。使用该注解完成属性注入时，类中无须 setter。当然，若属性有 setter，则也可将其加到 setter 上。（项目源码参见本书配套资源：第 5 章/基于注解的依赖注入/spring5）

（1）在项目 spring5 的 User 类中添加下面的注解：

```
@Scope("prototype")
@Component("user")
public class User {
    @Value("1")
    private String uid;
    @Value("李白")
    private String uname;
    @Value("男")
    private String gender;
    @Value("18")
    private int age;
    public void show(){
        System.out.println("用户编号:"+uid+" 用户姓名:"+uname+"性别:"+gender+" 年龄:"+age);
    }
//省略其他方法
}
```

（2）测试类 TestUser1 的代码如下：

```
public static void main(String[] args) {
        ApplicationContextcontext=newClassPathXmlApplicationContext ("application
```

```
Context.xml");
        User user=(User) context.getBean("user");
        user.show();
    }
```

测试结果如下：

用户编号:1 用户姓名:李白 性别:男 年龄:18

5.7.4 按类型注入域属性@Autowired

项目案例：使用注解，实现按类型注入域属性值。（项目源码参见本书配套资源：第 5 章/基于注解的依赖注入/spring6）

关键步骤：

（1）复制 spring5 为 spring6，先应用下面注解创建 Company 类的 Bean：

```
@Component("mycompany")
public class Company {
    private int id;
    @Value("砺锋科技")
    private String companyname;
    @Value("广州天河软件园")
    private String address;
    //省略 getter,setter 方法
    public String toString(){
        return "公司名称:"+collegename+" 公司地址:"+address;
    }
}
```

（2）在 User 类的 company 属性上添加@Autowired 注解，具体如下：

```
@Component("user")
public class User {
    @Value("1")
    private String uid;
    @Value("李白")
    private String uname;
    @Value("男")
    private String gender;
    @Value("18")
    private int age;
    @Autowired
    private Company company;
    //省略其他方法
}
```

默认是按类型装配，尽管属性名称是 company，Bean 的 id 是 mycompany，两者名称不一致，但由于是按类型装配，而不是按名称装配，所以可以注入。

（3）测试类 TestUser1 的代码如下：

```
public static void main(String[] args) {
    ApplicationContextcontext=newClassPathXmlApplicationContext ("application
Context.xml");
    User user=(User) context.getBean("user");
    user.show();
```

```
    System.out.println(user.getCompany());
}
```
测试结果如下：
用户编号:1 用户姓名:李白 性别:男 年龄:18
公司名称:砺锋科技，公司地址:广州天河软件园

5.7.5 按名称注入域属性@Autowired 与@Qualifier

需要在域属性上联合使用注解@Autowired 与@Qualifier。@Qualifier 的 value 属性用于指定要匹配的 Bean 的 id 值。

项目案例：使用注解，实现按名称注入域属性值。（项目源码参见本书配套资源：第5章/基于注解的依赖注入/spring6）

关键步骤：

（1）在项目 spring6 的 User 类中使用@Autowired 和@Qualifier 注解域属性，具体如下：

```
@Component("user")
public class Student {
    //省略其他代码
    @Autowired
    @Qualifier("mycompany")
    private Company company;
}
```

这时，域属性 company 必须装配 id 为 mycompany 的 Bean。如果找不到就会报错。

（2）测试类 TestUser1 的代码不变。测试结果如下：
用户编号:1 用户姓名:李白 性别:男 年龄:18
公司名称:砺锋科技，公司地址:广州天河软件园

（3）如果 User 类中修改@qualifier("mycompany")为@qualifier("mycompny")，再次运行测试类（注意，这里故意把"mycompany"打错为"mycompny"，少了一个字母）。

<u>Exceptioninthread"main"org.springframework.beans.factory. UnsatisfiedDependency
Exception</u>: Error creating bean with name 'user': Unsatisfied dependency
expressedthroughfield'company';nestedexception is <u>org.springframework.beans.factory.
NoSuchBeanDefinitionException</u>: No qualifying bean of type 'com.lifeng.entity.Company'
available: expected at least 1 bean which qualifies as autowire candidate. Dependency
annotations: {@org.springframework.beans.factory.annotation. Autowired(required=true),
@org.springframework.beans.factory.annotation. Qualifier(value=mycompany)}

结果系统报错，提示找不到匹配的 Bean。所以按名称注入域属性需要确保 Bean 的名称与注解一致。

@Autowired 还有一个属性 required，默认值为 true，表示匹配失败时会终止程序运行。若将 required 的值设置为 false，则匹配失败，将被忽略而不会报错，未匹配的属性值为 null。

（4）修改上一步的注解@Autowired 为@Autowired(required=false)，仍然使用拼写错误的@qualifier("mycompny")，代码如下：

```
@Component("user")
public class User {
    //省略其他代码
    @Autowired(required=false)
    @Qualifier("mycompny")
    private Company company;
```

再次运行测试类，结果如下：
用户编号:1 用户姓名:李白 性别:男 年龄:18
null

由测试得知，这次 Bean 名称虽然匹配不上，但并没有报错，只是域属性没有注入到值，显示为 null。

5.7.6 域属性注解@Resource

使用@Resource 注解既可以按名称匹配 Bean，也可以按类型匹配 Bean。（项目源码参见本书配套资源：第 5 章/基于注解的依赖注入/spring6）

（1）按类型注入域属性

@Resource 注解若不带任何参数，则会按照类型进行 Bean 的匹配注入，示例代码如下：

```
@Component("user")
public class User {
    //省略其他代码
    @Resource
    private Company company;
}
```

（2）按名称注入域属性。

@Resource 注解指定其 name 属性，则 name 的值即为按照名称进行匹配的 Bean 的 id，代码如下：

```
@Component("user")
public class User {
    //省略其他代码
    @Resource(name="mycompany")
    private Company company;
}
```

（3）测试结果同前。

5.7.7 XML 配置方式与注解方式的比较

XML 配置方式的优点：对其进行修改时，无须编译代码，只需重启服务器即可将新的配置加载。
注解方式的优点：配置方便，直观；缺点：以硬编码的方式写入 Java 代码中，若要修改，需要重新编译代码。

若注解方式与 XML 配置方式同用，则 XML 配置方式的优先级要高于注解方式。

上机练习

1. 使用 XML 配置方式，复制项目 spring4 为 springTest4，添加一个名为 Classes 的类，属性有班级编号、班级名称。修改 Student 类，添加 classes 属性。在配置文件时定义一个 id 为 classes 的 Bean，注入自定义的属性值，再在 id 为 student12 的 Bean 中引用这个 Bean。新建测试类 TestStudent8，输出学生的基本信息和所在班级的信息。

2. 需求同上，改为注解方式。

思考题

1. 基本类型如何注入属性值?
2. 集合类型如何注入属性值?
3. 域属性如何注入值?
4. 如何实现自动装配?
5. 注解有哪些?各有什么作用?

第 6 章　Spring 面向切面编程

本章目标

- 了解 AOP 编程的作用
- 了解切面、切点的概念
- 掌握使用 AspectJ 实现 AOP 的方法
- 掌握使用 XML 配置文件方式实现 AOP 的方法
- 掌握使用注解方式实现 AOP 的方法

AOP（Aspect Oriented Programming，面向切面编程）一般用于主业务需要切入系统业务的场合。如主业务需要记录日志，可在不改变主业务代码的情况下，切入日志功能。除了切入日志，AOP还可用于访问权限控制、事务管理、异常处理等系统级业务。传统编程模式下，系统级业务会频繁重复使用并与主业务代码混在一起，造成冗余及可维护性差，为了解决这个问题，把系统级业务定义为"切面"，把需要用到系统级业务功能的主业务方法定义为"切点"，通过配置使"切面"切入"切点"，这样当"切点"代表的方法执行时就会自动执行该"切面"，与拦截器的原理类似。

6.1 传统编程模式的弊端

学习下述案例，了解其中的弊端与解决问题的思路。

项目案例：在添加用户的同时输出日志。（项目源码参见本书配套资源：第6章/传统编程模式的弊端/spring7）

实现步骤：

（1）复制 spring6 为 spring7，在 src 下新建一个 log4j.properties 文件，以便输出日志，内容如下：

```
# Global logging configuration
log4j.rootLogger=ERROR, stdout
# MyBatis logging configuration...
log4j.logger.com.lifeng=DEBUG
# Console output...
log4j.appender.stdout=org.apache.log4j.ConsoleAppender
log4j.appender.stdout.layout=org.apache.log4j.PatternLayout
log4j.appender.stdout.layout.ConversionPattern=%5p [%t] - %m%n
```

如果不想用日志，可省略这一步，只需用 System.out.println()代替下面出现的 log.info()语句即可。

（2）Spring 配置文件 applicationContext.xml，代码如下：

```xml
<?xml version="1.0" encoding="UTF-8"?>
<beans xmlns="http://www.springframework.org/schema/beans"
    xmlns:xsi="http://www.w3.org/2001/XMLSchema-instance"
    xmlns:context="http://www.springframework.org/schema/context"
xsi:schemaLocation="
        http://www.springframework.org/schema/beans
http://www.springframework.org/schema/beans/spring-beans.xsd
        http://www.springframework.org/schema/context
http://www.springframework.org/schema/context/spring-context.xsd">
    <!-- bean definitions here -->
    <context:component-scan base-package="com.lifeng"/>
</beans>
```

（3）新建 com.lifeng.dao 包，包下新建 UserDao 接口、实现类 UserDaoImpl，使用注解@Component("userDao")将 UserDaoImpl 定义为一个 Bean：

```java
public interface UserDao {
    public void show();
    public void addUser();
}
@Component("userDao")
public class UserDaoImpl implements UserDao{
    @Override
```

```java
    public void show() {
        System.out.println("用户姓名:李白,所在公司:砺锋科技");
    }
    @Override
    public void addUser() {
        System.out.println("新增一个用户到数据库中");
    }
}
```

（4）新建 com.lifeng.service 包，包下新建类 UserService 里有 addUser()方法，完成新增学生的功能，使用注解@Component("userService")将 UserService 类定义为一个 Bean，参考代码如下：

```java
@Component("userService")
public class UserService {
    @Resource(name="userDao")
    private UserDao userDao;
    public void setUserDao(UserDao userDao) {
        this.userDao = userDao;
    }
    public void addUser(){
        userDao.addUser();
    }
    public void show(){
        userDao.show();
    }
}
```

如果想实现在 addUser 操作时同时记录日志，可修改代码如下：

```java
import javax.annotation.Resource;
import org.apache.log4j.Logger;
import org.springframework.stereotype.Component;
import com.lifeng.dao.UserDao;
@Component("userService")
public class UserService {
    private static final Logger log=Logger.getLogger(UserService.class);
    @Resource(name="userDao")
    private UserDao userDao;
    public void setUserDao(UserDao userDao) {
        this.userDao = userDao;
    }
    public void addUser(){
        log.info("开始添加用户...");
        userDao.addUser();
        log.info("完成添加用户...");
    }
    public void show(){
        userDao.show();
    }
}
```

（嵌入日志相关代码）

也就是说，在主业务代码的开头和结尾位置分别添加输出日志的相关代码。

（5）新建测试类 TestUser1，代码如下：

```java
public static void main(String[] args) {
    ApplicationContext context=new ClassPathXmlApplicationContext("applicationContext.xml");
```

```
    UserService userService=(UserService) context.getBean("userService");
    userService.addUser();
}
```

测试结果如下：

```
INFO [main] - 开始添加用户...
新增一个用户到数据库中
INFO [main] - 完成添加用户...
```

可以看到，虽然在调用 addUser()方法时实现了日志的输出，但这个程序有个问题，就是不得不在 addUser()方法的主业务代码前面添加 "log.info("开始添加用户...");"语句，以及在主业务代码后面添加 "log.info("开始添加用户...")"语句。这些语句以硬编码的方式混入主业务代码中，难以分割，可移植性差。另外，如果多种方法都要用到这些语句，则所有方法都要在头部和尾部反复编写这些代码，使程序变得更加复杂和冗余。如果要修改日志的格式，则会出现频繁而大量的修改。

解决问题的思路是将这些日志之类的功能独立出来，作为独立的一个或多个类，在需要时调用，而且最好不是显式调用，因为如果是显式调用，仍然要在目标方法的前面或后面嵌入代码，改进效果有限，最好是能自动调用。例如，在上面的案例中，addUser()方法只保留主业务代码，不混入任何有关日志的代码，但通过第三方的配置，让 addUser()方法执行时自动调用有关日志的功能，日志的功能本身也可单独成为一个类或多个类。

这种设想就是一种面向切面编程（AOP）的思想，我们可以把日志的功能想象成一个"切面"，切入到目标类的 addUser()方法的开始部位或结束部位，可把开始或结束部位想象为"切点"。

6.2 AOP 初试身手

下面先介绍有关 AOP 的一些基本概念。

（1）切面：一个单独的类，通常在此类中定义一些辅助功能或系统级功能的方法，如日志、权限管理、异常处理、事务处理等。这个类中的方法在需要时可以切入主业务方法（切点），可以切入主业务方法的前面、后面等位置。

（2）切点：主业务类的有些方法只想专注于完成核心业务逻辑，不想混入一些辅助性的功能，可以把这些方法定义为切点，切点就是确定什么位置放置切面。

（3）通知：切点是确定了使用切面的位置，但什么时候应用切面就由通知来决定，比如可以通知在切点方法执行前或切点方法执行后，或在出现异常时应用切面，这个时机称为通知。

（4）织入：切面和切点都是独立的功能类，通过织入才能让切面切入切点，所以织入就是一种配置过程，让切面能够精确地切入指定的位置。

项目案例：用面向切面的编程的方式改造上面这个项目。（项目源码参见本书配套资源：第 6 章/AOP 初试身手/spring8）

实现步骤：

（1）将项目 spring7 复制为 spring8，新添加的 JAR 包，如图 6.1 所示。

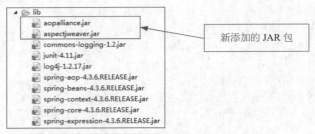

图 6.1　新添加的 JAR 包

（2）修改 UserService 类，删除有关日志功能的代码，只保留主业务代码，这个类的 addUser() 方法将作为切点，代码如下：

```
public void addUser(){
    log.info("开始添加用户....");
    userDao.addUser();
    log.info("完成添加用户....");
}
```

（3）新建名为 com.lifeng.aop 的包，包下新建一个名为 LoggerBefore 的类，这个类将作为一个切面，拟作为记录开头的日志，代码如下：

```
public class LoggerBefore implements MethodBeforeAdvice{
    private static final Logger log=Logger.getLogger(LoggerBefore.class);
    @Override
    public void before(Method arg0, Object[] arg1, Object arg2) throws Throwable {
        log.info("开始添加学生...");
    }
}
```

该方法实现了一个叫 MethodBeforeAdvice（前置通知）的接口，意味着将来这个方法会作用到切入目标的开始部位。

（4）在包 com.lifeng.aop 下新建一个名为 LoggerAfterReturning 的类，这个类将作为一个切面，拟作为记录结束位置的日志，代码如下：

```
public class LoggerAfterReturning implements AfterReturningAdvice{
    private static final Logger log=Logger.getLogger(LoggerAfterReturning.class);
    @Override
    public void afterReturning(Object arg0, Method arg1, Object[] arg2, Object arg3) throws Throwable {
        log.info("完成添加学生...");
    }
}
```

该方法实现了一个叫 AfterReturningAdvice（后置通知）的接口，意味着将来这个方法会作用到切入目标的结束部位。从上面几个类来看，主业务功能和辅助的日志功能，已用不同的类完全分开，不存在交叉。接下来配置，让主业务能调用到辅助的日志功能。

（5）修改配置文件。

首先要引入 AOP 约束，请参考 Spring 解压缩文件包中的文件：spring-framework-4.3.4.RELEASE-dist/spring-framework-4.3.4.RELEASE/docs/spring-framework-reference/html/xsd-configuration.html，如图 6.2 所示。

第6章 Spring 面向切面编程

```
§ 41.2.7 the aop schema
The aop tags deal with configuring all things AOP in Spring: this includes Spring's own proxy-based AOP framework and Spring's integration with the AspectJ AOP
framework. These tags are comprehensively covered in the chapter entitled Chapter 11, Aspect Oriented Programming with Spring.
In the interest of completeness, to use the tags in the aop schema, you need to have the following preamble at the top of your Spring XML configuration file; the text in
the following snippet references the correct schema so that the tags in the aop namespace are available to you.

<?xml version="1.0" encoding="UTF-8"?>
<beans xmlns="http://www.springframework.org/schema/beans"
    xmlns:xsi="http://www.w3.org/2001/XMLSchema-instance"
    xmlns:aop="http://www.springframework.org/schema/aop" xsi:schemaLocation="
        http://www.springframework.org/schema/beans http://www.springframework.org/schema/beans/spring-beans.xsd
        http://www.springframework.org/schema/aop http://www.springframework.org/schema/aop/spring-aop.xsd"> <!-- bean definitions here -->

</beans>
```

图 6.2　AOP 约束

```xml
<?xml version="1.0" encoding="UTF-8"?>
<beans xmlns="http://www.springframework.org/schema/beans"
   xmlns:xsi="http://www.w3.org/2001/XMLSchema-instance"
   xmlns:aop="http://www.springframework.org/schema/aop"     ← 引入 AOP
   xmlns:context="http://www.springframework.org/schema/context"
   xsi:schemaLocation="
     http://www.springframework.org/schema/beans
     http://www.springframework.org/schema/beans/spring-beans.xsd
     http://www.springframework.org/schema/context
     http://www.springframework.org/schema/context/spring-context.xsd
     http://www.springframework.org/schema/aop
     http://www.springframework.org/schema/aop/spring-aop.xsd">
   <context:component-scan base-package="com.lifeng"/>
   <!--定义切面-->
   <bean id="loggerBefore" class="com.lifeng.aop.LoggerBefore"/>          1. 定义切面
   <bean id="loggerAfterReturning" class="com.lifeng.aop.LoggerAfterReturning"/>
   <aop:config>
        <!-- 定义切点 -->                                2. 定义切点
        <aop:pointcut expression="execution(* com.lifeng.service.UserService.addUser())" id="pointcut"/>
        <!-- 通知切点,切入 advice-ref 指定的 Bean(切面)里面的方法,切入位置在前还是在后由切面的接口决定 -->
        <aop:advisor pointcut-ref="pointcut" advice-ref="loggerBefore"/>    3. 通知切点
        <aop:advisor pointcut-ref="pointcut" advice-ref="loggerAfterReturning"/>
   </aop:config>
</beans>
```

注意要引入 xmlns:aop=http://www.springframework.org/schema/aop 约束。基本流程是：先定义切面，再定义切点，最后通知切点切入切面，完成织入。

上述代码中，<aop:config>的配置是关键，正是它实现了在主业务的切点位置切入的功能（切面）；<aop:pointcut>子节点用于定义切点，可以用一个也可模糊匹配多个方法作为切点，id 是切点的名称，expression 则匹配方法，凡匹配上的方法都将设置为切点。本例中 expression="execution(*com.lifeng.service.UserService.addUser())"只匹配一个切点。就是 com.lifeng.service 包下的 UserService 类的 addUser()被定义为切点。

切点匹配规则的解释如下：

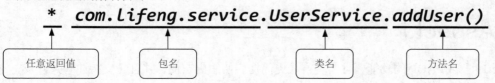

上面语句的意思是：将 com.lifeng.service 包下 UserService 类的 addUser()方法设置为切点。更多的情况下需要设置多个方法作为切点，所以要用到模糊匹配，常用的匹配手法如下所示：

上面语句的意思是：将 com.lifeng.service 包下的所有类的所有方法均设置为切点。

定义切点的常用方法如表 6.1 所示。

表 6.1 定义切点的常用办法

表达式	说明
execution(public * *(...))	指定切点为任意公共方法
execution(* save *(...))	指定切点为任何一个以"save"开始的方法
execution(* com.lifeng.service.*.*(...))	指定切点为定义在 service 包里的任意类的任意方法
execution(* *.service.*.doUpdate())	指定只有一级的包下的 serivce 子包下所有类中的 doUpdate()方法为切点
execution(* ..service.*. doUpdate ())	指定所有包下 serivce 子包下的所有类中的 doUpdate()方法为切点
execution(* com.lifeng.service.UserService.*(...))	指定切点为 UserService 类中的任意方法
execution(* com.lifeng.service.UsertService +.*(...))	指定切点：UserService 若为接口，则为接口中的任意方法及其所有实现类中的任意方法；若为类，则为该类及其子类中的任意方法
execution(*doRegister (String,*)))	指定切点为：所有的 doRegister ()方法，该方法的第 1 个参数为 String，第 2 个参数可以是任意类型，如 doRegister (String s1,String s2)和 doRegister (String s1,double d2)都匹配

（6）测试类的代码如下：
```
public static void main(String[] args) {
    ApplicationContextcontext=new ClassPathXmlApplicationContext("applicationContext.xml");
    UserService userService=(UserService) context.getBean("userService");
    userService.addUser();
}
```
测试结果如下：
INFO [main] - 开始添加学生...
新增一个学生到数据库中
 INFO [main] - 完成添加学生...

通过这个案例可以发现，主业务和辅助业务完全可以分开，可以将辅助业务想象为一个切面，将主业务想象为切点，通过 Spring 的配置将切面切入主业务的切点，这就是 AOP 编程的好处。这种编程也可理解为主业务的一种功能扩展，无须改动主业务类的代码，即可实现功能的扩展或增强。刚才的 LoggerBefore 类可看作是一种前置增强，或者叫前置通知；LoggerAfterReturning 类可看作是一种后置增强，或者叫后置通知。

6.3 AspectJ

AspectJ 可实现 AOP 的功能，且实现方式更简捷，使用更方便，并且还支持注解式开发，Spring

已将 AspectJ 对于 AOP 的实现也引入自己的框架中。

在 Spring 中使用 AOP 开发时，一般使用 AspectJ 的实现方式。AspectJ 是一个面向切面的框架，定义了 AOP 的语法。AspectJ 有一个专门的编译器用来生成遵守 Java 字节编码规范的 Class 文件。6.2 节的案例用的正是 AspectJ 的实现方式，项目中引入了 aspectjweaver.jar 包。

AspectJ 通过通知来完成切面切入切点（织入），根据 AspectJ 切点的位置和通知的时机，AspectJ 中常用的通知有 5 种类型。

（1）前置通知：MethodBeforeAdvice。
（2）后置通知：AfterReturningAdvice。
（3）环绕通知：MethodInterceptor。
（4）异常通知：ThrowsAdvice。
（5）最终通知：AfterAdvice。

其中，最终通知是指无论程序执行是否正常，该通知都会执行。类似于 Java SE 中 try...catch…finally 中的 finally 代码块。

除了上例提到的前置通知和后置通知外，下面介绍其他的通知类型，分别需要实现不同的接口。

6.3.1 异常通知

（1）在项目 spring8 的 com.lifeng.aop 包下新建一个名为 ErrorLogger 的类，实现 ThrowsAdvice 接口，代码如下：（项目源码参见本书配套资源：第 6 章/AOP 初试身手/spring8）

```java
public class ErrorLogger implements ThrowsAdvice{
    private static final Logger log=Logger.getLogger(ErrorLogger.class);
    public void afterThrowing(Method arg0, Object[] arg1, Object arg2,RuntimeException e){
        log.error(arg0.getName()+"这个方法发生了异常，异常信息："+e);
    }
}
```

（2）在 UserService 下新建一个方法：

```java
public void method(){
        int num=10/0;
        System.out.println("test!");
}
```

（3）修改配置文件，代码如下：

```xml
<?xml version="1.0" encoding="UTF-8"?>
<beans xmlns="http://www.springframework.org/schema/beans"
   xmlns:xsi="http://www.w3.org/2001/XMLSchema-instance"
   xmlns:aop="http://www.springframework.org/schema/aop"
   xmlns:context="http://www.springframework.org/schema/context"
   xsi:schemaLocation="
      http://www.springframework.org/schema/beans
      http://www.springframework.org/schema/beans/spring-beans.xsd
      http://www.springframework.org/schema/context
      http://www.springframework.org/schema/context/spring-context.xsd
      http://www.springframework.org/schema/aop
      http://www.springframework.org/schema/aop/spring-aop.xsd">
    <context:component-scan base-package="com.lifeng"/>
    <!--定义切面-->
    <bean id="errorLogger" class="com.lifeng.aop.ErrorLogger"/>
    <bean id="loggerBefore" class="com.lifeng.aop.LoggerBefore"/>
```

1.定义切面

```xml
        <bean id="loggerAfterReturning" class="com.lifeng.aop.LoggerAfterReturning"/>
        <aop:config>
            <!-- 定义切点 2 -->                                                       2.定义切点
            <aop:pointcutexpression="execution(* com.lifeng.service.UserService.method())" id="pointcut2"/>
            <!-- 定义切点 -->
            <aop:pointcutexpression="execution(* com.lifeng.service.UserService.addUser())" id="pointcut"/>
            <!-- 通知切点 2-->                                                        3.通知切点
            <aop:advisor pointcut-ref="pointcut2" advice-ref="errorLogger"/>
            <!-- 通知切点,切入 advice-ref 指定的 Bean(切面)里的方法,切入位置在前还是在后由切面的接口决定 -->
            <aop:advisor pointcut-ref="pointcut" advice-ref="loggerBefore"/>
            <aop:advisor pointcut-ref="pointcut" advice-ref="loggerAfterReturning"/>
        </aop:config>
</beans>
```

将 ErrorLogger 定义为一个切面，将 StudentService 类中的 method()方法定义为切点 pointcut1，使用通知实现切面 ErrorLogger 切入切点 method()。

由上面代码也可发现，可以有多个切面、切点，只要 id 不同就不会冲突。

（4）新建测试类 TestUser2 关键代码如下：

```java
public static void main(String[] args) {
    ApplicationContextcontext=new ClassPathXmlApplicationContext("applicationContext.xml");
    UserService userService =(UserService) context.getBean("userService");
    userService.addUser();
    userService.method();
}
```

（5）测试结果如下：

```
INFO [main] - 开始添加用户...
新增一个用户到数据库中
INFO [main] - 完成添加用户...
ERROR [main] - method 这个方法发生了异常,异常信息:
Exception in thread "main" java.lang.ArithmeticException: / by zero
```

可以看到除了原来的日志继续有效外，异常通知也起到了作用。

6.3.2　环绕通知

环绕通知可以在目标方法的前面后面都切入增强功能，类似拦截器，Spring 把控制权交给环绕通知，可以获取目标方法的参数、返回值，也可以进行异常处理、权限控制，还可以决定目标方法是否执行。

实现步骤：

（1）在 com.lifeng.aop 包下新建类 AroundLog，实现 MethodInterceptor 接口，将作为切面，代码如下。（项目源码参见本书配套资源：第 6 章/AOP 初试身手/spring8）

```java
public class AroundLog implements MethodInterceptor{
    private static final Logger log=Logger.getLogger(ErrorLogger.class);
    @Override
    public Object invoke(MethodInvocation invoke) throws Throwable {
```

```
            log.info(invoke.getMethod()+"方法执行之前...");
            Object result= invoke.proceed();//执行目标方法
            log.info(invoke.getMethod()+"方法执行之后...");
            return result;
        }
}
```

（2）在 userService 下新建一个方法 method2()，将其作为切点：
```
public void method2(){
    System.out.println("test 测试环绕通知!");
}
```

（3）修改配置文件，代码如下：
```xml
<?xml version="1.0" encoding="UTF-8"?>
<beans xmlns="http://www.springframework.org/schema/beans"
    xmlns:xsi="http://www.w3.org/2001/XMLSchema-instance"
    xmlns:aop="http://www.springframework.org/schema/aop"
    xmlns:context="http://www.springframework.org/schema/context"
    xsi:schemaLocation="
        http://www.springframework.org/schema/beans
        http://www.springframework.org/schema/beans/spring-beans.xsd
        http://www.springframework.org/schema/context
        http://www.springframework.org/schema/context/spring-context.xsd
        http://www.springframework.org/schema/aop
        http://www.springframework.org/schema/aop/spring-aop.xsd">
    <context:component-scan base-package="com.lifeng"/>
    <!--定义切面-->
    <bean id="aroundLogger" class="com.lifeng.aop.AroundLog"/>
    <bean id="errorLogger" class="com.lifeng.aop.ErrorLogger"/>
    <bean id="loggerBefore" class="com.lifeng.aop.LoggerBefore"/>
    <bean id="loggerAfterReturning" class="com.lifeng.aop.LoggerAfterReturning"/>
    <aop:config>
        <!-- 定义切点 2 -->
        <aop:pointcutexpression="execution(* com.lifeng.service.UserService.method())" id="pointcut2"/>
        <!-- 定义切点-->
        <aop:pointcutexpression="execution(* com.lifeng.service.UserService.addUser())" id="pointcut"/>
        <!-- 定义切点 3 -->
        <aop:pointcutexpression="execution(* com.lifeng.service.UserService.method2())" id="pointcut3"/>
        <!-- 通知切点 2-->
        <aop:advisor pointcut-ref="pointcut2" advice-ref="errorLogger"/>
        <!-- 通知切点,切入 advice-ref 指定的 Bean(切面)里面的方法,切入位置在前还是在后由切面的
接口决定 -->
        <aop:advisor pointcut-ref="pointcut" advice-ref="loggerBefore"/>
        <aop:advisor pointcut-ref="pointcut" advice-ref="loggerAfterReturning"/>
        <!-- 通知切点 3-->
        <aop:advisor pointcut-ref="pointcut3" advice-ref="aroundLogger"/>
    </aop:config>
</beans>
```

（4）新建测试类 TestUser3，关键代码如下：
```
public static void main(String[] args) {
        ApplicationContext context=new ClassPathXmlApplicationContext("application
```

```
Context.xml");
            UserService userService =(UserService) context.getBean("userService");
            userService.method2();
    }
```
测试结果如下：
```
INFO [main] - public void com.lifeng.service.UserService.method2()方法执行之前...
test 测试环绕通知！
 INFO [main] - public void com.lifeng.service.UserService.method2()方法执行之后...
```
上述代码简单地在目标方法的前面和后面输出了日志，相当于环绕着目标方法做了一些事情，功能上相当于同时实现了前置通知和后置通知。如果要进行权限控制，显然只要在调用 invkoe.proceed()执行目标方法之前进行就可以了，权限检查未通过的则不调用 invoke.proceed()方法，表示不执行目标方法。

6.4 使用注解实现通知

上述定义为切面的类（也可称为切面类）都是通过实现各种特定接口来实现的，缺点是必须实现特定接口，且一个类基本上只有一个重写方法是有用的。普通的类，不实现特定的接口，能否作为切面呢？对于普通的类不实现这些特定的接口，可以通过注解转化为通知类（切面类）。

首先在类上面添加@Aspect 注解，将该类转化为切面类，再在类中的各种方法上面使用各自的"通知"注解即可实现。各种"通知"注解作用如下：

- @Aspect 注解将此类定义为切面；
- @Before 注解用于将目标方法配置为前置增强（前置通知）；
- @AfterReturning 注解用于将目标方法配置为后置增强（后置通知）；
- @Around 定义环绕增强（环绕通知）；
- @AfterThrowing 配置异常通知；
- @After 也是后置通知，与@AfterReturning 相似，区别在于@AfterReturning 在方法执行完毕后返回，可以有返回值，而@After 没有返回值。

项目案例：将一个普通类用注解转化为切面类。（项目源码参见本书配套资源：第 6 章/使用注解实现通知/spring9）

实现步骤：

（1）将项目 spring8 复制为 spring9，在 com.lifeng.aop 包下新建一个名为 MyLog 的类，不实现任何接口，代码如下：
```
public class MyLog {
    private static final Logger log=Logger.getLogger(ErrorLogger.class);
    public void beforeMethod(){
        log.info("开始执行方法....");
    }
    public void afterMethod(){
        log.info("完成执行方法....");
    }
}
```

（2）在类 MyLog 中添加注解，将其转化为切面类，代码如下：
```
@Aspect
public class MyLog {
    private static final Logger log=Logger.getLogger(MyLog.class);
    @Before("execution(* com.lifeng.service.UserService.addUser())")
    public void beforeMethod(){
        log.info("开始执行方法...");
    }
    @AfterReturning("execution(* com.lifeng.service.UserService.addUser())")
    public void afterMethod(){
        log.info("完成执行方法...");
    }
}
```
execution 用于定义切点。

（3）清空原配置文件，修改如下：
```
<?xml version="1.0" encoding="UTF-8"?>
<beans xmlns="http://www.springframework.org/schema/beans"
    xmlns:xsi="http://www.w3.org/2001/XMLSchema-instance"
    xmlns:aop="http://www.springframework.org/schema/aop"
    xmlns:context="http://www.springframework.org/schema/context"
    xsi:schemaLocation="
       http://www.springframework.org/schema/beans
       http://www.springframework.org/schema/beans/spring-beans.xsd
       http://www.springframework.org/schema/context
       http://www.springframework.org/schema/context/spring-context.xsd
       http://www.springframework.org/schema/aop
       http://www.springframework.org/schema/aop/spring-aop.xsd">
   <context:component-scan base-package="com.lifeng"/>
   <aop:aspectj-autoproxy/>
    <bean id="myLog" class="com.lifeng.aop.MyLog"/>
```

上述代码中，关键是<aop:aspectj-autoproxy/>这条语句启动了对@AspectJ 注解的支持，还有添加了注解的类也要在配置文件中创建一个 Bean。具体到本案例，必须有<bean id="myLog" class="com.lifeng.aop.MyLog"/>这个配置。可以发现，使用了注解后，配置文件大大简化了。

（4）测试类 TestUser1 的代码如下：
```
public static void main(String[] args) {
        ApplicationContextcontext=new ClassPathXmlApplicationContext("application
Context.xml");
        UserService userService =(UserService) context.getBean("userService");
        userService.addUser();
    }
```
测试结果如下：
INFO [main] - 开始执行方法...
新增一个用户到数据库中
 INFO [main] - 完成执行方法...

6.5　使用 XML 定义切面

上述增强类，要么用到了特定接口，要么用到了注解，各有不足之处。使用注解有个问题，就是直接在类里面编写注解，不能分离，不方便修改与维护。其实还有一种办法，既不需要实现任何特定接口，也不需要用注解，只需在第三方的配置文件中配置一下，即可让普通的类具有增强类的效果，这就是使用 XML 配置文件的切面的方式，也是目前最常用的 AOP 方式。切面还可以获取切点方法的参数，也可不获取。下面介绍 XML 中配置 AspectJ 的若干标签的含义，如表 6.2 所示。

表 6.2　配置 AspectJ 标签

标签名称	解释
<aop:config>	配置 AOP，实现将一个普通类设置为切面类的功能
<aop:aspect>	配置切面
<aop:pointcut>	定义切点
<aop:before>	配置前置通知，将一个切面方法设置为前置增强
<aop:after-returning>	配置后置通知，将一个切面方法设置为后置增强，可有返回值
<aop:around>	配置环绕通知
<aop:after-throwing>	配置异常通知，将一个切面方法设置为异常抛出增强
<aop:after>	配置最终通知，将一个切面方法设置为后置增强，无返回值,无论是否异常都执行

6.5.1　切面不获取切点参数

　　项目案例：通过 XML 配置方式，让一个普通的类有前置、后置通知的效果。（项目源码参见本书配套资源：第 6 章/使用 xml 配置定义切面/spring10）

　　实现步骤：

　　（1）复制 spring9 为 spring10，在包 com.lifeng.aop 下新建一个名为 MyLogger 的普通类，该类不实现特定接口，也不加注解，代码如下：

```
public class MyLogger {
    private static final Logger log=Logger.getLogger(MyLogger.class);
    public void beforeMethod(){
        log.info("开始执行方法...");
    }
    public void afterMethod(){
        log.info("完成执行方法...");
    }
}
```

　　（2）清空原配置文件，修改为如下代码：

```xml
<?xml version="1.0" encoding="UTF-8"?>
<beans xmlns="http://www.springframework.org/schema/beans"
    xmlns:xsi="http://www.w3.org/2001/XMLSchema-instance"
    xmlns:aop="http://www.springframework.org/schema/aop"
    xmlns:context="http://www.springframework.org/schema/context"
    xsi:schemaLocation="
        http://www.springframework.org/schema/beans
        http://www.springframework.org/schema/beans/spring-beans.xsd
        http://www.springframework.org/schema/context
        http://www.springframework.org/schema/context/spring-context.xsd
        http://www.springframework.org/schema/aop
        http://www.springframework.org/schema/aop/spring-aop.xsd">
    <context:component-scan base-package="com.lifeng"/>
    <!-- 定义切面 -->                                          1. 定义切面
    <bean id="myLogger" class="com.lifeng.aop.MyLogger"/>
```

```xml
<aop:config>
    <!-- 定义切点 -->                                        2. 定义切点
    <aop:pointcut  expression="execution(*  com.lifeng.service.*.*(..))" id="pointcut"/>
    <!-- 引用包含增强方法的Bean(切面)，设置各种通知 -->      3. 通知切点
    <aop:aspect ref="myLogger">
        <!-- 将切面的beforeMethod方法设置为前置通知 -->
        <aop:before method="beforeMethod" pointcut-ref="pointcut"></aop:before>
        <!-- 将切面的afterMethod方法设置为后置通知，有返回值 -->
        <aop:after-returningmethod="afterMethod" pointcut-ref="pointcut"></aop:after-returning>
    </aop:aspect>
</aop:config>
</beans>
```

（3）测试类的关键代码如下：

```java
public static void main(String[] args) {
    ApplicationContextcontext=new ClassPathXmlApplicationContext("applicationContext.xml");
    UserService userService=(UserService) context.getBean("userService");
    userService.addUser();
}
```

测试结果如下：

```
INFO [main] - 开始执行方法...
新增一个用户到数据库中
 INFO [main] - 完成执行方法...
```

【注意】切入点有多个，可根据需要引用。

6.5.2 切面获取切点方法的参数与返回值

项目案例：将一个普通类设置为切面，同时能够获取切点方法的参数以及切点方法的返回值。（项目源码参见本书配套资源：第6章/使用xml配置定义切面/spring11）

实现步骤：

（1）将项目spring10复制为spring11，新建一个名为MyAspect的普通类，代码如下：

```java
public class MyAspect {
    public void beforeSave(String uname){
        System.out.println("开始执行添加...名为"+ uname +"的用户");
    }
    public void afterSave(String uname){
        System.out.println("完成执行添加...名为"+ uname +"的用户\n\n");
    }
    public void beforeUpdate(String uid,String uname,String gender,int age){
        System.out.println("开始执行修改...编号id为:"+ uid+",名为:"+ uname +",性别为:"+gender+",年龄为:"+age+"的用户");
    }
    public void afterUpdate(String uid,String uname,String gender,int age){
        System.out.println("完成执行修改...编号id为:"+ uid+",名为:"+ uname +",性别为: "+ gender +",年龄为:"+age+"的用户\n\n");
    }
    public void beforeDelete(){
```

```
            System.out.println("开始删除用户,返回 start deleting");
    }
    //切面获取到切点的返回值
    public void afterDelete(String result){
        System.out.println(result+"\n\n");
    }
    public void throwMethod(Throwable t){
        System.out.println("出现异常啦,异常信息:"+t.getMessage());
    }
}
```

（2）修改 UserService 类，添加几个方法，代码如下：

```
@Component("stuService")
public class StudentService {
    public StudentService(){
    }
    public void save(String uname){
        System.out.println("向数据库中保存用户信息...");
    }
    public void update(String uid,String uname,String gender,int age){
        System.out.println("向数据库中修改用户信息...");
    }
    public String delete(){
        System.out.println("向数据库中删除用户 信息...");
        return "删除成功返回 success";
    }
    public void methodThrow(){
        Integer.parseInt("a");
    }
}
```

（3）Spring 配置文件修改如下，代码如下：

```
<?xml version="1.0" encoding="UTF-8"?>
<beans xmlns="http://www.springframework.org/schema/beans"
    xmlns:xsi="http://www.w3.org/2001/XMLSchema-instance"
    xmlns:aop="http://www.springframework.org/schema/aop"
    xmlns:context="http://www.springframework.org/schema/context"
    xsi:schemaLocation="
        http://www.springframework.org/schema/beans
        http://www.springframework.org/schema/beans/spring-beans.xsd
        http://www.springframework.org/schema/context
        http://www.springframework.org/schema/context/spring-context.xsd
        http://www.springframework.org/schema/aop
        http://www.springframework.org/schema/aop/spring-aop.xsd">
    <context:component-scan base-package="com.lifeng"/>
    <!-- 定义切面 -->
    <bean id="myAspect" class="com.lifeng.aop.MyAspect"/>
    <aop:config>
        <!-- 定义切点 -->
        <aop:pointcutexpression="execution(* com.lifeng.service.UserService.save(String)) and args(uname)" id="pointcut1"/>
        <aop:pointcutexpression="execution(* com.lifeng.service.UserService.update(String,String,String,int)) and args(uid,uname,gender,age)" id="pointcut2"/>
        <aop:pointcutexpression="execution(* com.lifeng.service.UserService.delete())" id="pointcut3"/>
```

```xml
            <aop:pointcut expression="execution(* com.lifeng.service.UserService.*(..))" id="pointcut"/>
            <!--配置切面，引用切面的Bean -->
            <aop:aspect ref="myAspect">
                <aop:before method="beforeSave" pointcut-ref="pointcut1"/>
                <aop:after method="afterSave" pointcut-ref="pointcut1"/>

                <aop:before method="beforeUpdate" pointcut-ref="pointcut2"/>
                <aop:after method="afterUpdate" pointcut-ref="pointcut2"/>

                <aop:before method="beforeDelete" pointcut-ref="pointcut3"/>
                <aop:after-returning  method="afterDelete"  pointcut-ref="pointcut3" returning="result"/>
                <aop:after-throwing  method="throwMethod"  pointcut-ref="pointcut" throwing="t"/>
            </aop:aspect>
        </aop:config>
    </beans>
```

【注意】expression="execution(* com.lifeng.service.UserService.save(String)) and args(uname)"在定义切点的同时声明了该切点方法的形式参数，用于传递给切面，该参数名称几个地方保持一致（分别是切点、切面、配置文件，都用同一个名称）。

<aop:after-returning method="afterDelete" pointcut-ref="pointcut3" returning="result"/>这句话在通知节点的同时声明了用于返回值的参数，该参数名称与切面类中 afterDelete 方法的参数名称也要一致，这样切面就可获取到切点方法中的返回值。

（4）测试类 TestUser1，关键代码如下：

```
public static void main(String[] args) {
        ApplicationContextcontext=new ClassPathXmlApplicationContext("application Context.xml");
        UserService userService=(UserService) context.getBean("userService");
        userService.save("张三");
        userService.update("1", "李四", "女", 18);
        userService.delete();
        userService.methodThrow();
    }
```

测试结果如下：
开始执行添加...名为张三的用户
向数据库中保存用户信息...
完成执行添加...名为张三的用户

开始执行修改...编号id为:1,名为:李四,性别为: 女,年龄为:18 的用户
向数据库中修改用户信息...
完成执行修改...编号id为:1,名为:李四,性别为: 女,年龄为:18 的用户

开始删除用户,返回 start deleting
向数据库中删除用户 信息...
删除成功返回 success

出现异常啦,异常信息:For input string: "a"
Exception in thread "main" java.lang.NumberFormatException: For input string: "a"

可见切面获取了切点方法的参数，甚至还获取了切点方法的返回值。

上机练习

复制项目 spring16，在 StudentService 类中添加 Method01()、Method02(String myname)和 Method03()，再在 com.lifeng.aop 包下新建一个名为 MyApsect2 的类，用来做切面，在该类中定义 beforMehtod01、afterMethod01、beforeMethod02、afterMethod02、afterReurningMethod02、afterReurningMethod03。其中，beforeMethod02、afterMethod02 要获取并输出切点的参数，afterMethod02 要能输出返回值。

思考题

1. 通知类型有哪些?
2. 切面如何获取切点的参数?

第 7 章　Spring 操作数据库

本章目标

 ✧　掌握 Spring 模板类 JdbcTemplate 的各种方法
 ✧　学会各种数据源的配置方法

使用原始 JDBC 操作数据的代码比较复杂、冗长，因此 Spring 专门提供了一个模板类 JdbcTemplate 来简化 JDBC 的操作。JdbcTemplate 提供了 execute 方法、update 方法、query 方法等，各个方法又有多种重载或变形可以满足常用的增删改查，如表 7.1 所示。

表 7.1 JdbcTemplate 的各种方法

方法	描述
execute(String sql)	用于执行 SQL 语句，可以是任何 SQL 语句，但通常只是 DDL 语句
int update(String sql)	根据 SQL 中的不带参数的增删改语句执行数据操作，返回受影响的行数
Int update(String sql,Object…args)	这个 SQL 通常是带 "?" 占位符参数的，args 用于为这些占位符参数赋值，同样返回受影响的行数
List<T>query(String sql, RowMapper<T> rowMapper)	执行不带参数的 select 语句，返回多条值的情况，封装成 T 类型的泛型集合，事先要定义好 RowMapper<T>对象 rowMapper
List<T>query(Stringsql,Object[] args, RowMapper<T> rowMapper)	执行带参数的 select 语句，返回多条值的情况，封装成 T 类型的泛型集合，事先要定义好 RowMapper<T>对象 rowMapper
SqlRowSet queryForRowSet(String sql)	可用于查询部分列，或者类似 count(*) 的聚合查询语句，返回 SqlRowSet 行集合。需要调用 next 方法移到行集合的第一行，再用 getInt(列号)获取
TqueryForObject(Stringsql, RowMapper<T> rowMapper)	执行不带参数的 select 语句，返回单条值的情况，封装成 T 类型，事先要定义好 RowMapper<T>对象 rowMapper
TqueryForObject(String sql,Object[] args,RowMapper<T> rowMapper)	执行带参数的 select 语句，返回单条值的情况，封装成 T 类型，事先要定义好 RowMapper<T>对象 rowMapper

7.1 JdbcTemplate 数据源

要使用 JdbcTemplate，首先要在 Spring 配置文件中配置数据源 dataSource 连接到数据库，再配置 JdbcTemplate 模板，将 dataSource 注入给 Jdbctemplate 的 dataSource 属性，最后将 JdbcTemplate 实例化对象注入给调用者的 jdbcTemplate 属性，调用者即可使用。有以下三种数据源可供选择：

① Spring 默认的数据源 DriverManagerDataSource；
② DBCP 数据源；
③ C3P0 数据源。

7.1.1 DriverManagerDataSource 数据源

DriverManagerDataSource 数据源的 Spring 配置文件示例如下：

```
<?xml version="1.0" encoding="UTF-8"?>
<beans xmlns="http://www.springframework.org/schema/beans"
    xmlns:xsi="http://www.w3.org/2001/XMLSchema-instance"
    xmlns:aop="http://www.springframework.org/schema/aop"
    xmlns:context="http://www.springframework.org/schema/context"
    xsi:schemaLocation="
       http://www.springframework.org/schema/beans
       http://www.springframework.org/schema/beans/spring-beans.xsd
       http://www.springframework.org/schema/context
       http://www.springframework.org/schema/context/spring-context.xsd
       http://www.springframework.org/schema/aop
       http://www.springframework.org/schema/aop/spring-aop.xsd">
    <!--1.配置数据源,连接数据库 -->
```

```xml
<bean id="dataSource" class="org.springframework.jdbc.datasource.DriverManagerDataSource">
        <property name="driverClassName">
            <value>com.mysql.jdbc.Driver</value>
        </property>
        <property name="url">
            <value>jdbc:mysql://localhost:3306/studentdb?characterEncoding=utf8</value>
        </property>
        <property name="username">
            <value>root</value>
        </property>
        <property name="password">
            <value>root</value>
        </property>
    </bean>
    <!--2. 配置 jdbcTemplate 模板 注入 dataSource -->
    <bean id="jdbcTemplate" class="org.springframework.jdbc.core.JdbcTemplate">
        <property name="dataSource" ref="dataSource" />
    </bean>
    <!--3. 配置 DAO,注入 jdbcTemplate 属性值 -->
    <bean id="userDao" class="com.lifeng.dao.UserDaoImpl">
        <property name="jdbcTemplate" ref="jdbcTemplate"/>
    </bean>
</beans>
```

7.1.2 DBCP 数据源 BasicDataSource

使用 DBCP 数据源 BasicDataSource 可以实现连接池的功能，需要导入以下两个 JAR 包：commons-dbcp-osgi-1.2.2.jar 和 commons-pool-1.5.3.jar。

Spring 配置文件示例如下：

```xml
<?xml version="1.0" encoding="UTF-8"?>
<beans xmlns="http://www.springframework.org/schema/beans"
    xmlns:xsi="http://www.w3.org/2001/XMLSchema-instance"
    xmlns:aop="http://www.springframework.org/schema/aop"
    xmlns:context="http://www.springframework.org/schema/context"
    xsi:schemaLocation="
        http://www.springframework.org/schema/beans
        http://www.springframework.org/schema/beans/spring-beans.xsd
        http://www.springframework.org/schema/context
        http://www.springframework.org/schema/context/spring-context.xsd
        http://www.springframework.org/schema/aop
        http://www.springframework.org/schema/aop/spring-aop.xsd">
    <!-- 配置数据源 -->
    <bean id="dataSource" class="org.apache.commons.dbcp.BasicDataSource">
        <property name="driverClassName">
            <value>com.mysql.jdbc.Driver</value>
        </property>
        <property name="url">
            <value>jdbc:mysql://localhost:3306/studentdb?characterEncoding=utf8</value>
        </property>
        <property name="username">
            <value>root</value>
        </property>
```

这里改用了 DBCP 数据源

```xml
        <property name="password">
            <value>root</value>
        </property>
    </bean>
    <!-- 省略其他代码-->
</beans>
```

7.1.3 C3P0 数据源 ComboPooledDataSource

同样能实现连接池的功能，需要导入 JAR 包 c3p0-0.9.1.2.jar，并且配置关键字也有不同之处，详见下面的示例代码：（项目源码参见本书配套资源：第 7 章/ c3p0 数据源/spring1702）

```xml
    <!-- 配置数据源 -->
    <bean id="dataSource" class="com.mchange.v2.c3p0.ComboPooledDataSource">
        <property name="driverClass">
            <value>com.mysql.jdbc.Driver</value>
        </property>
        <property name="jdbcUrl">
            <value>jdbc:mysql://localhost:3306/studentdb?characterEncoding=utf8</value>
        </property>
        <property name="user">
            <value>root</value>
        </property>
        <property name="password">
            <value>root</value>
        </property>
    </bean>
    <!-- 省略其他代码-->
```

注意不同之处

7.1.4 使用属性文件读取数据库连接信息

能否把数据库连接信息单独放在一个属性文件中，与配置文件分开呢？若能，则更换数据库时只需要修改该属性文件，便于数据库连接信息的修改与维护。（项目源码参见本书配套资源：第 7 章/使用属性文件/spring1703）

关键步骤：

（1）在 src 下添加属性文件 jdbc.properties，代码如下，其中包含了数据库连接所需要的 4 条信息：

```
jdbc.driver=com.mysql.jdbc.Driver
jdbc.url=jdbc:mysql://localhost:3306/studentdb
jdbc.username=root
jdbc.password=root
```

（2）修改配置文件，先要注册属性文件再在数据源配置中用 EL 表达式引用属性文件，代码如下：

```xml
<?xml version="1.0" encoding="UTF-8"?>
<beans xmlns="http://www.springframework.org/schema/beans"
    xmlns:xsi="http://www.w3.org/2001/XMLSchema-instance"
    xmlns:aop="http://www.springframework.org/schema/aop"
    xmlns:context="http://www.springframework.org/schema/context"
    xsi:schemaLocation="
        http://www.springframework.org/schema/beans
        http://www.springframework.org/schema/beans/spring-beans.xsd
        http://www.springframework.org/schema/context
        http://www.springframework.org/schema/context/spring-context.xsd
        http://www.springframework.org/schema/aop
```

```xml
             http://www.springframework.org/schema/aop/spring-aop.xsd">
    <!-- 注册属性文件的第一种方式 -->
    <bean class="org.springframework.beans.factory.config.PropertyPlaceholderConfigurer">
        <property name="location" value="classpath:jdbc.properties"/>
    </bean>
    <!-- 配置数据源 -->
    <bean id="dataSource" class="com.mchange.v2.c3p0.ComboPooledDataSource">
        <property name="driverClass">
            <value>${jdbc.driver}</value>
        </property>
        <property name="jdbcUrl">
            <value>${jdbc.url}</value>
        </property>
        <property name="user">
            <value>${jdbc.username}</value>
        </property>
        <property name="password">
            <value>${jdbc.password}</value>
        </property>
    </bean>
    <!-- 省略其他代码 -->
</beans>
```

要引用数据库属性文件需要注册该属性文件，注册属性文件除了上面用到第一种方式，还有另一种方式，代码如下：

```xml
<!-- 注册属性文件的第二种方式 -->
<context:property-placeholder location="classpath:jdbc.properties"/>
```

这种方式要求配置文件头中有 context 约束信息。这两种效果相同，选用其中一个即可。考虑到第二种方法简单些，本案例选用第二种。

7.2 JdbcTemplate 方法的应用

项目案例：使用表 7.1 的 JdbcTemplate 的各种方法操作数据库 User，实现增删改查数据。（项目源码参见本书配套资源：第 7 章/ JdbcTemplate 方法的应用/spring12）

实现步骤：

（1）新建项目 spring12，导入 JAR 包，包结构如图 7.1 所示。

图 7.1 所需 JAR 包

特别要注意多了以下 3 个 JAR 包：spring-jdbc-4.3.4.RELEASE.jar、spring-tx-4.3.4.RELEASE.jar、mysql-connector-java-5.1.37.jar。

（2）创建 MySQL 数据库 usersdb。

（3）新建名为 com.lifeng.entity 的包，包下创建实体类 User，代码如下：

```java
public class User {
    private String uid;
    private String uname;
    private String gender;
    private int age;
    public User(){
    }
    public User(String uid,String uname,String gender,int age){
        this.uid=uid;
        this.uname=uname;
        this.gender=gender;
        this.age=age;
    }
    public void show(){
        System.out.println("用户编号:"+uid+" 用户姓名:"+uname+" 用户性别:"+gender+" 用户年龄:"+age);
    }
    //省略 setter,getter 方法
}
```

（4）新建 com.lifeng.dao 包，包下创建 UserDao 接口，代码如下：

```java
public interface UserDao {
    public void create();
    public void add(User user);
    public void delete(int id);
    public void update(User user);
    public List<User> findAllUsers();
    public User findUserById(int id);
    public int count();
}
```

（5）创建 UserDao 接口的实现类 UserDaoImpl，代码如下：

```java
package com.lifeng.dao;
import java.util.List;

import org.springframework.jdbc.core.BeanPropertyRowMapper;
import org.springframework.jdbc.core.JdbcTemplate;
import org.springframework.jdbc.core.RowMapper;
import org.springframework.jdbc.support.rowset.SqlRowSet;
import com.lifeng.entity.User;
public class UserDaoImpl implements UserDao{
    JdbcTemplate jdbcTemplate;
    public JdbcTemplate getJdbcTemplate() {
        return jdbcTemplate;
    }
    public void setJdbcTemplate(JdbcTemplate jdbcTemplate) {
        this.jdbcTemplate = jdbcTemplate;
    }
//创建数据库表
@Override
```

```java
    public void create() {
        String sql="create table user(uid varchar(20),uname varchar(50),gender varchar(6),age int);";
        jdbcTemplate.execute(sql);
    }
    @Override
    public void add(User user) {
        String sql="insert into user(uid,uname,gender,age) values(?,?,?,?)";
        Object[] params={ user.getUid(),user.getUname(),user.getGender(),user.getAge()};
        jdbcTemplate.update(sql,params);
    }
    @Override
    public void delete(int id) {
        String sql="delete from user where uid=?";
        Object[] params={id};
        jdbcTemplate.update(sql,params);
    }
    @Override
    public void update(User user) {
        String sql="update user set uname=?,gender=?,age=? where uid=?";
        Object[] params={user.getUname(),user.getGender(),user.getAge(),user.getUid()};
        jdbcTemplate.update(sql,params);
    }
    @Override
    public List<User> findAllUsers() {
        String sql="select * from user";
        RowMapper<User> rowMapper=new BeanPropertyRowMapper<User>(User.class);
        List<User> list= jdbcTemplate.query(sql, rowMapper);
        return list;
    }
    @Override
    public User findUserById(int id) {
        String sql="select * from user where uid=?";
        Object[] params={id};
        RowMapper<User> rowMapper=new BeanPropertyRowMapper<User>(User.class);
        User user= jdbcTemplate.queryForObject(sql, params,rowMapper);
        return user;
    }
    @Override
    public int count() {
        String sql="select count(*) from user";
        SqlRowSet rs=jdbcTemplate.queryForRowSet(sql);
        if (rs.next()) {
            return rs.getInt(1);
        }
        return 0;
    }
}
```

上面代码定义了 jdbcTemplate 类型的 jdbcTemplate 对象属性，通过这个对象可以方便地操作数据库，无须再用以前的 JDBC 的烦琐的步骤，但这个对象目前还没实例化，显然等待下一步在配置文件中注入。

（6）配置文件 applicationContext.xml，代码如下：

```xml
<?xml version="1.0" encoding="UTF-8"?>
<beans xmlns="http://www.springframework.org/schema/beans"
    xmlns:xsi="http://www.w3.org/2001/XMLSchema-instance"
    xmlns:aop="http://www.springframework.org/schema/aop"
    xmlns:context="http://www.springframework.org/schema/context"
    xsi:schemaLocation="
        http://www.springframework.org/schema/beans
        http://www.springframework.org/schema/beans/spring-beans.xsd
        http://www.springframework.org/schema/context
        http://www.springframework.org/schema/context/spring-context.xsd
        http://www.springframework.org/schema/aop
        http://www.springframework.org/schema/aop/spring-aop.xsd">
    <!-- 配置数据源 -->
    <bean id="dataSource" class="org.springframework.jdbc.datasource.DriverManagerDataSource">
        <property name="driverClassName">
            <value>com.mysql.jdbc.Driver</value>
        </property>
        <property name="url">
            <value>jdbc:mysql://localhost:3306/usersdb </value>
        </property>
        <property name="username">
            <value>root</value>
        </property>
        <property name="password">
            <value>root</value>
        </property>
    </bean>
    <!-- 配置jdbcTemplate模板 注入dataSource -->
    <bean id="jdbcTemplate" class="org.springframework.jdbc.core.JdbcTemplate">
        <property name="dataSource" ref="dataSource" />
    </bean>
    <!-- 配置DAO,注入jdbcTemplate属性值 -->
    <bean id="userDao" class="com.lifeng.dao.UserDaoImpl">
        <property name="jdbcTemplate" ref="jdbcTemplate"/>
    </bean>
</beans>
```

这里先定义了一个名为 dataSource 的 Bean 作为数据源，里面配置了连接数据的所有信息。再定义了一个名为 jdbcTemplate 的 Bean，将上面定义的 dataSource 这个 Bean 注入该 Bean 的 dataSource 属性。最后定义一个名为 userDao 的 Bean，并将 jdbcTemplate 这个 Bean 注入该 Bean 的 jdbcTemplate 属性。

（7）测试类 TestUser1，关键代码如下：

```java
public class TestStudent1 {
    public static void main(String[] args) {
        ApplicationContext context=new ClassPathXmlApplicationContext("applicationContext.xml");
        UserDao userDao = (UserDao) context.getBean("userDao");
        System.out.println("----------创建用户数据表---------");
        userDao.create();
        System.out.println("----------创建用户表user 成功!---------");
```

```java
            System.out.println("\n----------添加若干个用户----------");
            User user = new User("1", "张飞", "男", 25);
            userDao.add(user);
            user = new User("2", "李白", "男", 20);
            userDao.add(user);
            user = new User("3", "李寻欢", "男", 22);
            userDao.add(user);
            user = new User("4", "赵敏", "女", 18);
            userDao.add(user);
            System.out.println("\n----------查找所有用户----------");
            List<User> list = userDao.findAllUsers();
            for (User u : list) {
                u.show();
            }
            System.out.println("\n----------查找一个用户----------");
            user = userDao.findUserById(1);
            user.show();
            System.out.println("\n----------修改一个用户----------");
            user = userDao.findUserById(3);
            user.setUname("小李飞刀");
            userDao.update(user);
            list = userDao.findAllUsers();
            for (User u : list) {
                u.show();
            }
            System.out.println("\n----------统计用户人数----------");
            System.out.println("用户总人数是:"+userDao.count());
            System.out.println("\n----------删除一个用户----------");
            userDao.delete(3);
            list = userDao.findAllUsers();
            for (User u : list) {
                u.show();
            }
        }
}
```

测试结果如下：
----------创建用户数据表----------
----------创建用户表user成功!----------
----------添加若干个用户----------
----------查找所有用户----------
用户编号:1 用户姓名:张飞 用户性别:男 用户年龄:25
用户编号:2 用户姓名:李白 用户性别:男 用户年龄:20
用户编号:3 用户姓名:李寻欢 用户性别:男 用户年龄:22
用户编号:4 用户姓名:赵敏 用户性别:女 用户年龄:18
----------查找一个用户----------
用户编号:1 用户姓名:张飞 用户性别:男 用户年龄:25
----------修改一个用户----------
用户编号:1 用户姓名:张飞 用户性别:男 用户年龄:25
用户编号:2 用户姓名:李白 用户性别:男 用户年龄:20

用户编号:3 用户姓名:小李飞刀 用户性别:男 用户年龄:22
用户编号:4 用户姓名:赵敏 用户性别:女 用户年龄:18
----------统计用户人数---------
用户总人数是:4
----------删除一个用户---------
用户编号:1 用户姓名:张飞 用户性别:男 用户年龄:25
用户编号:2 用户姓名:李白 用户性别:男 用户年龄:20
用户编号:4 用户姓名:赵敏 用户性别:女 用户年龄:18

上机练习

题目：连接第 1 章的数据库，操作商品表，利用 JdbcTemplate 进行增删改查操作。

思考题

1. 简述使用 JdbcTemplate 的步骤。
2. JdbcTemplate 有哪些常用方法？
3. 配置数据源的方法有哪几种？

第 8 章　Spring 事务管理

本章目标

- 了解事务隔离级别
- 掌握通过配置 XML 实现 Spring 事务管理的方法
- 掌握使用注解实现 Spring 事务管理的方法

事务（Transaction）是访问数据库的一个操作序列。这些操作要么都做，要么都不做，是一个不可分割的工作单元。通过事务，数据库能将逻辑相关的一组操作绑定在一起，以便保持数据的完整性。

事务有 4 个重要特性，简称 ACID。

（1）原子性（Automicity，A）：即事务中的所有操作要么全部执行，要么全部不执行。

（2）一致性（Consistency，C）：事务执行的结果必须是使数据库从一个一致状态变到另一个一致状态。

（3）隔离性（Isolation，I）：即一个事务的执行不能被另一个事务影响。

（4）持久性（Durabillity，D）：即事务提交后将被永久保存。

在 Java EE 开发中，事务原本属于 DAO 层的范畴，但一般情况下需要将事务提升到业务层(Service 层)，以便能够使用事务的特性来管理具体的业务。

8.1 Spring 事务管理接口

Spring 的事务管理会用到以下两个事务相关的接口。

8.1.1 事务管理器接口 PlatformTransactionManager

事务管理器接口 PlatformTransactionManager 主要用于完成对事务的提交、回滚操作，并能够获取事务的状态信息。PlatformTransactionManager 接口有两个常用的实现类。

（1）DataSourceTransactionManager 实现类：在通过 JDBC 或 MyBatis 进行数据持久化时使用。

（2）HibernateTransactionManager 实现类：在通过 Hibernate 进行数据持久化时使用。

关于 Spring 的事务提交与回滚方式，默认是：发生运行异常时回滚，发生受检查异常时提交，也就是说，程序抛出 runtime 异常的时候才会进行回滚，其他异常不回滚。

8.1.2 事务定义接口 TransactionDefinition

事务定义接口 TransactionDefinition 中定义了事务描述相关的 3 类常量：事务隔离级别常量、事务传播行为常量、事务默认超时时限常量，以及对它们的操作。

1. 事务隔离级别常量

在应用程序中，多个事务并发运行，操作相同的数据，可能会引起脏读、不可重复读、幻读等问题。

（1）脏读（Dirty read）：第一个事务访问并改写了数据，尚未提交事务，这时第二个事务进来了，读取了刚刚改写的数据，如果这时第一个事务回滚了，这样第二个事务读取到的数据就是无效的"脏数据"。

（2）不可重复读（Nonrepeatable read）：第一个事务在其生命周期内多次查询同一个数据，在两次查询之间，第二个事务访问并改写了该数据，导致第一个事务两次查询同一个数据得到的结果不一样。

（3）幻读（Phantom read）：幻读与不可重复读类似。它发生在第一个事务在其生命周期进行了两次按同一查询条件查询数据,第一次按该查询条件读取了几行数据,这时第二个事务进来了,且插入或删除了一些数据,然后第一个事务再次按同一条件查询,发现多了一些原本不存在的记录或者原有记录不见了。

为了解决并发问题,TransactionDefinition 接口定义了 5 个事务隔离常量如下:

① ISOLATION_DEFAULT:采用数据库默认的事务隔离级别。不同数据库不一样,MySql 的默认为 REPEATABLE_READ(可重复读);Oracle 默认为 READ_COMMITTED(读已提交)。

② ISOLATION_READ_UNCOMMITTED:读未提交。允许另外一个事务读取到当前事务未提交的数据,隔离级别最低,未解决任何并发问题,会产生脏读,不可重复读和幻像读。

③ ISOLATION_READ_COMMITTED:读已提交,被一个事务修改的数据提交后才能被另外一个事务读取,另外一个事务不能读取该事务未提交的数据。解决脏读,但还存在不可重复读与幻读。

④ ISOLATION_REPEATABLE_READ:可重复读。解决了脏读、不可重复读,但还存在幻读。

⑤ ISOLATION_SERIALIZABLE:串行化。按时间顺序一一执行多个事务,不存在并发问题,最可靠,但性能与效率最低。

2. 事务传播行为常量

事务传播行为是指处于不同事务中的方法在相互调用时,执行期间事务的维护情况。例如,当一个事务方法 B 调用另一个事务方法 A 时,应当明确规定事务如何传播,例如可以规定 A 方法继续在 B 方法的现有事务中运行,也可以规定 A 方法开启一个新事务,在新事务中运行,现有事务先挂起,等 A 方法的新事务执行完毕后再恢复。TransactionDefinition 接口一共定义了 7 种传播行为常量,说明如下:

(1)PROPAGATION_REQUIRED:指定的方法必须在事务内执行。若当前存在事务,就加入到当前事务中;若当前没有事务,则创建一个新事务。这种传播行为是最常见的选择,也是 Spring 默认的事务传播行为。如该传播行为加在 actionB()方法上,该方法将被 actionA()调用,若 actionA()方法在执行时就是在事务内的,则 actionB()方法的执行也加入到该事务内执行。若 actionA()方法没有在事务内执行,则 actionB()方法会创建一个事务,并在其中执行。

(2)PROPAGATION_SUPPORTS:指定的方法支持当前事务,但若当前没有事务,也可以以非事务方式执行。

(3)PROPAGATION_MANDATORY:指定的方法必须在当前事务内执行,若当前没有事务,则直接抛出异常。

(4)PROPAGATION_REQUIRES_NEW:总是新建一个事务,若当前存在事务,就将当前事务挂起,直到新事务执行完毕。

(5)PROPAGATION_NOT_SUPPORTED:指定的方法不能在事务环境中执行,若当前存在事务,就将当前事务挂起。

(6)PROPAGATION_NEVER:指定的方法不能在事务环境下执行,若当前存在事务,就直接抛出异常。

(7)PROPAGATION_NESTED:指定的方法必须在事务内执行。若当前存在事务,则在嵌套事务内执行;若当前没有事务,则创建一个新事务。

3. 默认事务超时时限

常量 TIMEOUT_DEFAULT 定义了事务底层默认的超时时限,及不支持事务超时时限设置的 none 值。该值一般使用默认值即可。

8.2 Spring 事务管理的实现方法

Spring 支持编程式事务和声明式事务,编程式事务直接在主业务代码中精确定义事务的边界,事务以硬编码的方式嵌入到了主业务代码里面。好处是能提供更加详细的事务管理,但由于编程式事务主业务与事务代码混在一起,不易分离,耦合度高,不利于维护与重用。声明式事务则基于 AOP 方式,将主业务操作与事务规则进行解耦,能在不影响业务代码具体实现的情况下实现事务管理。所以比较常用的是声明式事务。声明式事务又有两种具体的实现方式:基于 XML 配置文件的方式和基于注解的方式。

8.2.1 没有事务管理的情况分析

项目案例:模拟支付宝转账,张三、李四原本各有账户余额 2000 元,张三转账 500 元给李四,但转账过程中出现了异常。(项目源码参见本书配套资源:第 8 章/没有事务管理的情况/spring13)

实现步骤:

(1)复制项目 spring12 为 spring13,在 MySQL 中创建数据库表,代码如下:

```
create table alipay (
    aliname varchar (60),
    amount double
);
insert into alipay (aliname, amount) values('张三','2000');
insert into alipay (aliname, amount) values('李四','2000');
```

(2)在 com.lifeng.dao 包下创建 IAccountDao 接口,代码如下:

```
public interface AlipayDao {
    public void transfer(String fromA,String toB,int amount);
}
```

(3)在 com.lifeng.dao 包下创建 IAlipayDao 接口的实现类 AlipayDaoImpl,代码如下:

```
package com.lifeng.dao;
import org.springframework.jdbc.core.JdbcTemplate;
public class AlipayDaoImpl implements AlipayDao{
    JdbcTemplate jdbcTemplate;
    public JdbcTemplate getJdbcTemplate() {
        return jdbcTemplate;
    }
    public void setJdbcTemplate(JdbcTemplate jdbcTemplate) {
        this.jdbcTemplate = jdbcTemplate;
    }
    @Override
    public void transfer(String fromA, String toB, int amount) {
        jdbcTemplate.update("updatealipaysetamount=amount-?where aliname=?",amount,fromA);
        Integer.parseInt("a");
        jdbcTemplate.update("updatealipaysetamount=amount+?where aliname=?",amount,
```

```
            toB);
        }
    }
```
这个 transfer 方法主要实现两个操作。操作一：转出操作，张三的账户减少钱；操作二：转入操作，李四的账户增加钱，但两个操作中间模拟出了差错（异常）。这将导致张三的钱减少了，李四的钱却没增加。

（4）修改 Spring 配置文件，代码如下：
```xml
<?xml version="1.0" encoding="UTF-8"?>
<beans xmlns="http://www.springframework.org/schema/beans"
    xmlns:xsi="http://www.w3.org/2001/XMLSchema-instance"
    xmlns:aop="http://www.springframework.org/schema/aop"
    xmlns:context="http://www.springframework.org/schema/context"
    xsi:schemaLocation="
        http://www.springframework.org/schema/beans
        http://www.springframework.org/schema/beans/spring-beans.xsd
        http://www.springframework.org/schema/context
        http://www.springframework.org/schema/context/spring-context.xsd
        http://www.springframework.org/schema/aop
        http://www.springframework.org/schema/aop/spring-aop.xsd">
    <!-- 配置数据源 -->
    <beanid="dataSource" class="org.springframework.jdbc.datasource.DriverManagerDataSource">
        <property name="driverClassName">
            <value>com.mysql.jdbc.Driver</value>
        </property>
        <property name="url">
            <value>jdbc:mysql://localhost:3306/usersdb</value>
        </property>
        <property name="username">
            <value>root</value>
        </property>
        <property name="password">
            <value>root</value>
        </property>
    </bean>
    <!-- 配置 jdbcTemplate 模板 -->
    <bean id="jdbcTemplate" class="org.springframework.jdbc.core.JdbcTemplate">
        <property name="dataSource" ref="dataSource" />
    </bean>
    <!-- 配置 DAO,注入 jdbcTemplate 属性值 -->
    <bean id="alipayDao" class="com.lifeng.dao.AlipayDaoImpl">
        <property name="jdbcTemplate" ref="jdbcTemplate"/>
    </bean>       ← 新加的 Bean

    <!-- 配置 DAO 层,注入 jdbcTemplate 属性值 -->
    <bean id="userDao" class="com.lifeng.dao.UserDaoImpl">
        <property name="jdbcTemplate" ref="jdbcTemplate"/>
    </bean>
</beans>
```
（5）测试类 TestAlipay。
```
public class TestAlipay {
    public static void main(String[] args) {
```

```
            ApplicationContextcontext=new ClassPathXmlApplicationContext("application
Context.xml");
            AlipayDao alipayDao=(AlipayDao) context.getBean("alipayDao");
            alipayDao.transfer("张三", "李四", 500);
        }
    }
```

测试结果如下。

① 转账前的数据库，如图 8.1 所示。

② 转账后的数据库，如图 8.2 所示。

	aliname	amount
□	张三	2000
□	李四	2000
*	(NULL)	(NULL)

图 8.1 转账前的数据库

	aliname	amount
□	张三	1500
□	李四	2000
*	(NULL)	(NULL)

图 8.2 转账后的数据库

观察图 8.2 可发现，张三的 500 元转出去了，但李四的 500 元却没收到，看控制台的输出，系统提示发生了异常，如图 8.3 所示。

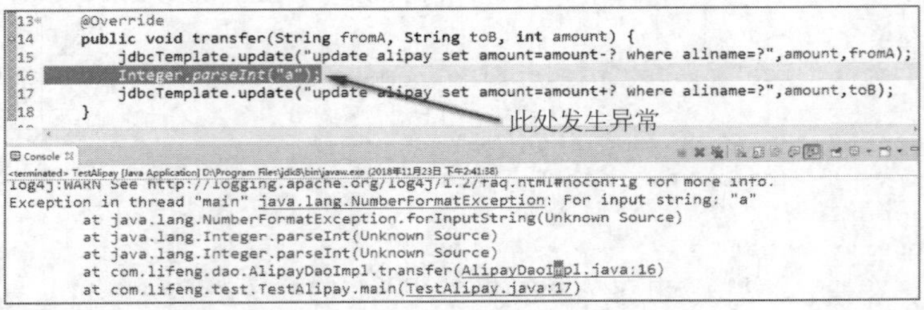

图 8.3 异常信息

上述程序在张三的钱刚转走的时候发生了异常，程序中断，而李四的钱来不及转进去了。这样就出现了问题。

解决问题的思路在于，转入/转出这两个操作应合计作为同一个事务，即要么同时成功，要么同时失败，不能只成功一半。

8.2.2 通过配置 XML 实现事务管理

下面进行事务管理方面的改进，目标是把类 AlipayDaoImpl 里的整个 transfer()方法作为事务管理，这样 transfer()里的所有操作（包括转出/转入操作）都纳入同一个事务，从而使 transfer()里的所有操作要么一起成功，要么一起失败。这里利用了 Spring 的事务管理机制进行处理。

项目案例：模拟支付宝转账，张三、李四原本各有账户余额 2000 元，张三转账 500 元给李四，但转账过程中间出现异常，导致数据不一致，现应用 Spring 的事务管理，配置 XML，避免不一致的情况。（项目源码参见本书配套资源：第 8 章/使用 xml 实现事务管理/spring14）

实现步骤：

（1）复制项目 spring13 为 spring14，修改配置文件，添加文件头约束，添加事务管理模块。

其中约束信息参考如下：

Spring 解压文件夹/spring-framework-4.3.4.RELEASE/docs/spring-framework-reference/ html/xsd-configuration.html，如图 8.4 所示。

> **41.2.6 the tx (transaction) schema**
>
> The `tx` tags deal with configuring all of those beans in Spring's comprehensive support for transactions. These tags are covered in the chapter entitled Chapter 17, Transaction Management.
>
> You are strongly encouraged to look at the `'spring-tx.xsd'` file that ships with the Spring distribution. This file is (of course), the XML Schema for Spring's transaction configuration, and covers all of the various tags in the `tx` namespace, including attribute defaults and suchlike. This file is documented inline, and thus the information is not repeated here in the interests of adhering to the DRY (Don't Repeat Yourself) principle.
>
> In the interest of completeness, to use the tags in the `tx` schema, you need to have the following preamble at the top of your Spring XML configuration file; the text in the following snippet references the correct schema so that the tags in the `tx` namespace are available to you.
>
> ```xml
> <?xml version="1.0" encoding="UTF-8"?>
> <beans xmlns="http://www.springframework.org/schema/beans"
>     xmlns:xsi="http://www.w3.org/2001/XMLSchema-instance"
>     xmlns:aop="http://www.springframework.org/schema/aop"
>     xmlns:tx="http://www.springframework.org/schema/tx" xsi:schemaLocation="
>     http://www.springframework.org/schema/beans http://www.springframework.org/schema/beans/spring-beans.xsd
>     http://www.springframework.org/schema/tx http://www.springframework.org/schema/tx/spring-tx.xsd
>     http://www.springframework.org/schema/aop http://www.springframework.org/schema/aop/spring-aop.xsd"> <!-- bean definitions here -->
> </beans>
> ```

图 8.4　约束信息

```xml
<?xml version="1.0" encoding="UTF-8"?>
<beans xmlns="http://www.springframework.org/schema/beans"
    xmlns:xsi="http://www.w3.org/2001/XMLSchema-instance"
    xmlns:aop="http://www.springframework.org/schema/aop"
    xmlns:tx="http://www.springframework.org/schema/tx"
    xmlns:context="http://www.springframework.org/schema/context"
    xsi:schemaLocation="
        http://www.springframework.org/schema/beans
        http://www.springframework.org/schema/beans/spring-beans.xsd
        http://www.springframework.org/schema/context
        http://www.springframework.org/schema/context/spring-context.xsd
        http://www.springframework.org/schema/tx
        http://www.springframework.org/schema/tx/spring-tx.xsd
        http://www.springframework.org/schema/aop
        http://www.springframework.org/schema/aop/spring-aop.xsd">
    <!-- 配置数据源 -->
    <bean id="dataSource" class="org.springframework.jdbc.datasource.DriverManagerDataSource">
        <property name="driverClassName">
            <value>com.mysql.jdbc.Driver</value>
        </property>
        <property name="url">
            <value>jdbc:mysql://localhost:3306/usersdb </value>
        </property>
        <property name="username">
            <value>root</value>
        </property>
        <property name="password">
            <value>root</value>
        </property>
    </bean>
    <!-- 配置jdbcTemplate模板 注入dataSource -->
    <bean id="jdbcTemplate" class="org.springframework.jdbc.core.JdbcTemplate">
        <property name="dataSource" ref="dataSource" />
    </bean>
```

```xml
<!-- 配置DAO,注入jdbcTemplate属性值 -->
<bean id="alipayDao" class="com.lifeng.dao.AlipayDaoImpl">
    <property name="jdbcTemplate" ref="jdbcTemplate"/>
</bean>

<!-- 定义事务管理器 -->
<bean id="txManager"
    class="org.springframework.jdbc.datasource.DataSourceTransactionManager">
    <property name="dataSource" ref="dataSource" />
</bean>
<!-- 编写事务通知 -->
<tx:advice id="txAdvice" transaction-manager="txManager">
    <tx:attributes>
        <tx:method name="*" propagation="REQUIRED" isolation="DEFAULT" read-only="false" />
    </tx:attributes>
</tx:advice>
<!-- 编写AOP,让Spring自动将事务切入到目标切点 -->
<aop:config>
    <!-- 定义切入点 -->
    <aop:pointcut id="txPointcut"
        expression="execution(* com.lifeng.dao.*.*(..))" />
    <!-- 将事务通知与切入点组合 -->
    <aop:advisor advice-ref="txAdvice" pointcut-ref="txPointcut" />
</aop:config>
</beans>
```

这里可以把事务功能理解为切面，通过AOP配置实现事务（切面）自动切入到切入点（目标方法），从而将目标方法（切入点）纳入事务管理，而目标方法本身可以不用管事务，专心做自己的主业务功能就行了。

（2）其他程序不变，运行测试。

测试时尽管转账中间出现了异常，但张三、李四的钱都没变化，保持了一致性，这样就达到了目的，证明了 transfer 方法中的两个操作都纳入了同一个事务。发生异常时，事务回滚，保证了数据的一致性。

上述配置中的代码如下：
```xml
<tx:method name="*" propagation="REQUIRED"
        isolation="DEFAULT"     read-only="false" />
```

表示匹配的切点方法都进行事务管理，这里*表示匹配所有切点方法，propagation="REQUIRED"表示匹配的切点方法必须在事务内执行，isolation="DEFAULT"表示事务隔离级别默认，对于MySQL数据库，隔离级别为REPEATABLE_READ（可重复读）。read-only="false"表示非只读。

上述的配置粒度太大，所有方法都使用同一种事务管理模式，要想不同的方法实现不一样的事务管理，还得细化配置。项目中常见的细化配置如下面代码所示：

```xml
<!-- 编写通知 -->
<tx:advice id="txAdvice" transaction-manager="txManager">
    <tx:attributes>
        <tx:method name="save*" propagation="REQUIRED" />
        <tx:method name="add*" propagation="REQUIRED" />
        <tx:method name="insert*" propagation="REQUIRED" />
        <tx:method name="delete*" propagation="REQUIRED" />
        <tx:method name="update*" propagation="REQUIRED" />
```

```xml
            <tx:method name="search*" propagation="SUPPORTS" read-only="true"/>
            <tx:method name="select*" propagation="SUPPORTS" read-only="true"/>
            <tx:method name="find*" propagation="SUPPORTS" read-only="true"/>
            <tx:method name="get*" propagation="SUPPORTS" read-only="true"/>
        </tx:attributes>
    </tx:advice>
```

这样，不同的方法匹配不同的事务管理模式。

<tx:method name="save*" propagation="REQUIRED" />表示凡是以 save 开头的切点方法必须在事务内执行，其他增删改都一样的意思。对于查询操作，则使用 <tx:method name="select*" propagation="SUPPORTS" read-only="true"/>表示以 select 开头的切点方法支持当前事务，若当前没有事务，也可以非事务方式执行，read-only="true"表示只读，其他几个类似。

8.2.3 利用注解实现事务管理

前边介绍了利用 XML 配置文件实现事务管理的办法，下面介绍用注解实现事务管理。

在类或方法上使用@Transactional 注解，即可实现事务管理。@Transactional 注解有下面这些属性（可选）。

（1）propagation：用于设置事务传播的属性，该属性类型为 propagation 枚举，默认值为 Propagation.REQUIRED。

（2）isolation：用于设置事务的隔离级别，该属性类型为 Isolation 枚举，默认值为 Isolation.DEFAULT。

（3）readOnly：用于设置该方法对数据库的操作是否是只读的，该属性为 boolean，默认值 false。

（4）timeout：用于设置本操作与数据库连接的超时时限。单位为秒，类型为 int，默认值为-1，即没有时限。

（5）rollbackFor：指定需要回滚的异常类，类型为 Class[]，默认值为空数组。当然，若只有一个异常类时，可以不使用数组。

（6）rollbackForClassName：指定需要回滚的异常类的类名，类型为 String[]，默认值为空数组。当然，若只有一个异常类时，可以不使用数组。

（7）noRollbackFor：指定不需要回滚的异常类。类型为 Class[]，默认值为空数组。当然，若只有一个异常类时，可以不使用数组。

（8）noRollbackForClassName：指定不需要回滚的异常类类名。类型为 String[]，默认值为空数组。当然，若只有一个异常类时，可以不使用数组。

需要注意的是，@Transactional 若用在方法上，只能用在 public 方法上。对于其他非 public 方法，如果加上了注解@Transactional，虽然 Spring 不会报错，但不会将指定事务织入到该方法中。因为 Spring 会忽略掉所有非 public 方法上的@Transaction 注解。若@Transaction 注解在类上，则表示该类上的所有方法均将在执行时织入事务。

项目案例：模拟支付宝转账，张三、李四原本各有账户余额 2000 元，张三转账 500 元给李四，但转账过程中间出现异常，应用 Spring 的事务管理，使用注解，避免不一致的情况。（项目源码参见本书配套资源：第 8 章/使用注解实现事务管理/spring15）

实现步骤:

(1) 将项目 spring14 复制为 spring15,修改配置文件如下:

```xml
<?xml version="1.0" encoding="UTF-8"?>
<beans xmlns="http://www.springframework.org/schema/beans"
    xmlns:xsi="http://www.w3.org/2001/XMLSchema-instance"
    xmlns:aop="http://www.springframework.org/schema/aop"
    xmlns:tx="http://www.springframework.org/schema/tx"
    xmlns:context="http://www.springframework.org/schema/context"
    xsi:schemaLocation="
        http://www.springframework.org/schema/beans
        http://www.springframework.org/schema/beans/spring-beans.xsd
        http://www.springframework.org/schema/context
        http://www.springframework.org/schema/context/spring-context.xsd
        http://www.springframework.org/schema/tx
        http://www.springframework.org/schema/tx/spring-tx.xsd
        http://www.springframework.org/schema/aop
        http://www.springframework.org/schema/aop/spring-aop.xsd">
    <!-- 配置数据源 -->
    <bean id="dataSource" class="org.springframework.jdbc.datasource.DriverManagerDataSource">
        <property name="driverClassName">
            <value>com.mysql.jdbc.Driver</value>
        </property>
        <property name="url">
            <value>jdbc:mysql://localhost:3306/usersdb </value>
        </property>
        <property name="username">
            <value>root</value>
        </property>
        <property name="password">
            <value>root</value>
        </property>
    </bean>
    <!-- 配置jdbcTemplate模板 注入dataSource -->
    <bean id="jdbcTemplate" class="org.springframework.jdbc.core.JdbcTemplate">
        <property name="dataSource" ref="dataSource" />
    </bean>

    <!-- 配置DAO,注入jdbcTemplate属性值 -->
    <bean id="alipayDao" class="com.lifeng.dao.AlipayDaoImpl">
        <property name="jdbcTemplate" ref="jdbcTemplate"/>
    </bean>

    <!-- 定义事务管理器 -->
    <bean id="txManager"
        class="org.springframework.jdbc.datasource.DataSourceTransactionManager">
        <property name="dataSource" ref="dataSource" />
    </bean>
    <!-- 开启事务注解驱动 -->
    <tx:annotation-driven transaction-manager="txManager"/></beans>
```

可以发现,配置文件比之前简化了很多,事务方面,只须定义好事务管理器,再开启事务注解驱动即可。其他的交给注解来解决。

（2）利用@Transactional 注解修改转账方法。

@Transactional 既可以修饰类，也可以修饰方法，如果修饰类，则表示事务的设置对整个类的所有方法都起作用，如果修饰在方法上，则只对该方法起作用，代码如下：

```java
package com.lifeng.dao;
import org.springframework.jdbc.core.JdbcTemplate;
import org.springframework.transaction.annotation.Isolation;
import org.springframework.transaction.annotation.Propagation;
import org.springframework.transaction.annotation.Transactional;
public class AlipayDaoImpl implements AlipayDao{
    JdbcTemplate jdbcTemplate;
    public JdbcTemplate getJdbcTemplate() {
        return jdbcTemplate;
    }
    public void setJdbcTemplate(JdbcTemplate jdbcTemplate) {
        this.jdbcTemplate = jdbcTemplate;
    }

    @Override
    @Transactional(propagation=Propagation.REQUIRED,isolation=Isolation.DEFAULT,readOnly=false)
    public void transfer(String fromA, String toB, int amount) {
        jdbcTemplate.update("update alipay set amount=amount-? where aliname=?",amount,fromA);
        Integer.parseInt("a");
        jdbcTemplate.update("update alipay set amount=amount+? where aliname=?",amount,toB);
    }
}
```

（注解事务）

上述代码将 transfer()方法注解为事务。

（3）运行测试，发现数据库同样没改变，所以注解事务起到作用了。

8.2.4　在业务层实现事务管理

上面的案例是在 DAO 层实现事务管理，相对简单一些，但实际开发时需要在业务层实现事务管理，而不是在 DAO 层，为此，项目修改如下，特别要注意在业务层的事务管理实现。

项目案例：模拟支付宝转账，张三、李四原本各有账户余额 2000 元，张三转账 500 元给李四，但转账过程中间出现异常，在业务层应用 Spring 的事务管理，配置 XML，避免不一致的情况。（项目源码参见本书配套资源：第 8 章/在业务层实现事务管理/spring16）

实现步骤：

（1）复制项目 spring14 为 spring16，修改 DAO 层，将转出、转入分拆成两个方法，代码如下：

```java
public class AlipayDaoImpl implements AlipayDao{
    JdbcTemplate jdbcTemplate;
    public JdbcTemplate getJdbcTemplate() {
        return jdbcTemplate;
    }
    public void setJdbcTemplate(JdbcTemplate jdbcTemplate) {
        this.jdbcTemplate = jdbcTemplate;
    }
    @Override
    public void tranferFrom(String fromA,int amount){
```

```
            jdbcTemplate.update("update alipay set amount=amount-? where aliname=?",
amount,fromA);
        }
        @Override
        public void tranferTo(String toB,int amount){
            jdbcTemplate.update("updat ealipayset amount=amount+? where aliname=?",
amount,toB);
        }
    }
```

（2）新建包 com.lifeng.service，创建业务层 AlipayService.java 类，代码如下：
```
public class AlipayService {
    private AlipayDao alipayDao;
    public AlipayDao getAlipayDao() {
        return alipayDao;
    }
    public void setAlipayDao(AlipayDao alipayDao) {
        this.alipayDao = alipayDao;
    }
    public void transfer(String fromA, String toB, int amount) {
        alipayDao.tranferFrom(fromA, amount);
        Integer.parseInt("a");
        alipayDao.tranferTo(toB, amount);
    }
}
```

上述代码相当于把有异常问题的 transfer()方法迁移到业务层中来。

（3）修改配置文件。关键配置如下：
```
<!-- 配置 jdbcTemplate 模板 注入 dataSource -->
<bean id="jdbcTemplate" class="org.springframework.jdbc.core.JdbcTemplate">
    <property name="dataSource" ref="dataSource" />
</bean>
<!-- 配置 DAO,注入 jdbcTemplate 属性值 -->
<bean id="alipayDao" class="com.lifeng.dao.AlipayDaoImpl">
    <property name="jdbcTemplate" ref="jdbcTemplate"/>
</bean>
<!-- 配置 SERVICE 层,注入 alipayDao 属性值 -->
<bean id="alipayService" class="com.lifeng.service.AlipayService">
    <property name="alipayDao" ref="alipayDao"/>
</bean>
<!-- 定义事务管理器 -->
<bean id="txManager"
    class="org.springframework.jdbc.datasource.DataSourceTransactionManager">
    <property name="dataSource" ref="dataSource" />
</bean>
<!-- 编写事务通知 -->
<tx:advice id="txAdvice" transaction-manager="txManager">
    <tx:attributes>
        <tx:method name="*" propagation="REQUIRED"  isolation="DEFAULT"
read-only="false" />
    </tx:attributes>
</tx:advice>
<!-- 编写 AOP,让 Spring 自动将事务切入到目标切点 -->
<aop:config>
    <!-- 定义切入点 -->
```

```xml
        <aop:pointcut id="txPointcut"
            expression="execution(* com.lifeng.service.*.*(..))" />
        <!-- 将事务通知与切入点组合 -->
        <aop:advisor advice-ref="txAdvice" pointcut-ref="txPointcut" />
    </aop:config>
```
（4）修改测试类，代码如下：
```java
public class TestAlipay {
    public static void main(String[] args) {
        ApplicationContext context=new ClassPathXmlApplicationContext("applicationContext.xml");
        AlipayService alipayService=(AlipayService) context.getBean("alipayService");
        alipayService.transfer("张三", "李四", 500);
    }
}
```
测试结束，数据库的数据保持不变，证明事务管理成功。

上机练习

复制项目 spring13 为 spring13Test，分别使用 XML 配置方式和注解方式，实现业务层的 transfer 方法的事务管理。

思考题

1. 事务隔离级别有哪些？
2. 哪些方式可以实现事务管理？各自的实现过程是怎样的？

第 9 章　Spring MVC 入门

本章目标

- ✧ 了解 Spring MVC 基本原理
- ✧ 掌握 Spring MVC 开发环境的搭建方法

9.1 Spring MVC 简介

Spring MVC 是一种基于 MVC 设计模式的使用请求-响应模型的轻量级 Web 框架。MVC 是一种 Web 开发领域的设计模式，是 Model、View 与 Controller 三个英文单词的首字母缩写，指的是对 Web 应用程序中的资源按功能划分的 3 大部分。

（1）View（视图）：是用户进行操作的可视化界面，可以是 HTML、jsp、XML 等。用户可以在视图上看到服务端传来的数据，或者在视图上录入数据以便传递给服务端处理。

（2）Model（模型）：用于处理业务逻辑、封装、传输业务数据。

（3）Controller（控制器）：是程序的调度中心，控制程序的流转，接收客户端的请求，判断该调用哪个服务端程序来处理，处理完毕后把获得的模型数据显示到视图，返回给用户。

通过以上分工，将使程序更加简单高效。Struts2 和 Spring MVC 都是 MVC 框架。

9.1.1 Spring MVC 的优点

Spring MVC 与 Struts2 相比，具有更好的安全性、可靠性、运行速度更快。目前 Spring MVC 已成为 Java Web 开发的一款利器，越来越受到 Java 开发者的喜爱。Spring MVC 具有以下优点。

（1）角色划分清晰：核心控制器（DispatcherServlet）、处理器映射器（HandlerMapping）、处理器适配器（HandlerAdapter）、视图解析器（ViewResolver）、处理器（Controller）、验证器（Validator）、命令对象（Command，请求参数绑定的对象）、表单对象（Form Object，提供给表单展示和提交的对象）。

（2）分工明确，扩展灵活。作为 Spring 的一部分，易与 Spring 其他框架集成。

（3）可适配性好，通过 HandlerAdapter 就可以支持任意一个类作为处理器。

（4）支持数据验证、数据格式化、数据绑定机制。

（5）提供功能强大的 JSP 标签库，使数据在视图中的展示或者获取更加丰富与灵活。

（6）RESTful 风格的支持，文件的上传、下载功能简单。

（7）注解的零配置支持。

9.1.2 Spring MVC 的运行原理

Spring MVC 的运行原理和流程如图 9.1 所示。

Spring MVC 工作流程如下。

（1）浏览器向服务端提交请求，请求会被核心控制器 DispatcherServlet 拦截。

（2）核心控制器将请求转给处理器映射器 HandlerMapping。

（3）处理器映射器 HandlerMapping 会根据请求，找到处理该请求的具体的处理器，并将其封装为处理器执行链后返回给核心控制器 DispatcherServlet。

（4）核心控制器根据处理器执行链中的处理器，找到能够执行该处理器的处理器适配器 HandlerAdapter。

（5）处理器适配器 HandlerAdapter 调用执行处理器 Controller。

（6）处理器 Controller 将处理结果及要跳转的视图封装到一个对象 ModelAndView 中，并将其返回给处理器适配器 HandlerAdapter。

（7）处理器适配器 HandlerAdapter 直接将结果返回给核心控制器。

（8）核心控制器调用视图解析器 ViewResolver，将 ModelAndView 中的视图名称封装为视图对象 View。

（9）视图解析器 ViewResolver 将封装了的视图对象 View 返回给核心控制器 DispatcherServlet，到此一个流程结束。

（10）核心控制器 DispatcherServlet 调用视图对象 View，让其自己进行数据填充，形成响应对象。

（11）核心控制器把填充好数据的 View 响应给浏览器。

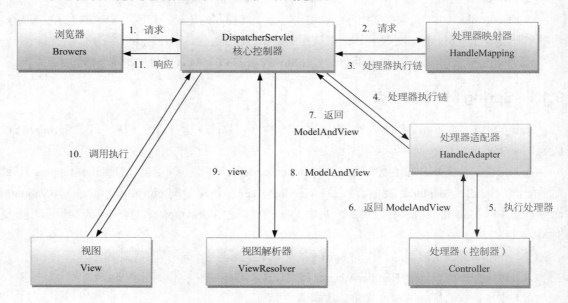

图 9.1 Spring MVC 运行原理和流程

9.2 第一个 Spring MVC 程序

9.2.1 开发环境

本书的案例使用了以下软件版本：JDK 1.8、Tomcat 8.5、Eclipse 4.6.3。

以上是 Java Web 开发的基本环境，相关软件请自行下载安装，在此基础上，要使用 Spring MVC 还需以下组件：Spring 4.3.4、Apache Commons Logging。

可以从 Spring 4.3.4 的官方网站下载 spring-framework-4.3.4.RELEASE-dist.zip，解压缩结构如图 9.2 所示。

其中 libs 文件夹下有很多 jar 包备用。从 Apache Commons Loging 组件的官方下载地址中下载 commons-logging-1.2-bin.zip，解压后可见到 commons-logging-1.2.jar，备用。其他的软件自行下载并安装。下面通过第一个 Spring MVC 程序来学习如何搭建 Spring MVC 框架开发环境。

第 9 章 Spring MVC 入门

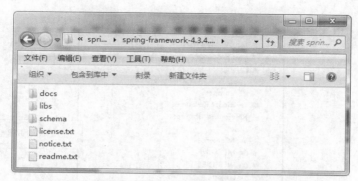

图 9.2 spring-framework-4.3.4.RELEASE-dist.zip 解压结构

9.2.2 第一个 Spring MVC 程序

项目案例：用户通过浏览器提交一个请求，服务端处理器在接收到这个请求后，给出一条欢迎信息"Hello Spring MVC"，并在响应页面中显示该信息，采用传统的配置式开发方式。（项目源码参见本书配套资源：第 9 章/第一个 mvc 程序/springmvc1）

实现步骤：

（1）在 Eclipse 中新建一个 Dynamic Web Project 项目。

新建项目命名为 springmvc1，在 src 下新建一个包，将其命名为 com.lifeng.controller 用于放置控制器类，在 WebContent 的 WEB-INF 下新建一个文件夹 jsp，用于放置视图，项目初始结构如图 9.3 所示。

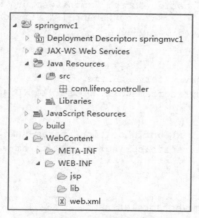

图 9.3 项目初始结构

（2）导入 Spring 的 JAR 包。

在项目的 WebConten/WEB-INF/lib 目录下添加 JAR 包，其中 spring 开头的 JAR 包在 libs 文件夹下，commons-logging-1.1.jar 包可直接下载。添加 JAR 包的方法是：复制所需的 JAR 包，然后粘贴到 WebContent/WEB-INF/lib 目录下。全部添加完后，再在 WebContent/WEB-INF/lib 目录下全选这些 JAR 包，单击鼠标右键，选择 add to build path 即可。添加完毕后如图 9.4 所示。

145

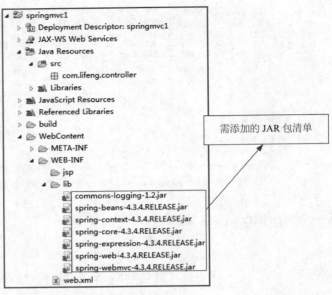

图9.4 需添加的 JAR 包

以上 JAR 包也可在本章配套源码中找到。

（3）在 web.xml 上配置 DispatcherServlet 核心控制器，项目 WebConten/WEB-INF 目录里的 web.xml 文件配置如下：

```xml
<?xml version="1.0" encoding="UTF-8"?>
<web-appxmlns:xsi="http://www.w3.org/2001/XMLSchema-instance" xmlns="http://xmlns.jcp.org/xml/ns/javaee" xsi:schemaLocation="http://xmlns.jcp.org/xml/ns/javaee http://xmlns.jcp.org/xml/ns/javaee/web-app_3_1.xsd" id="WebApp_ID" version="3.1">
    <display-name>springmvc1</display-name>
    <!-- 以下代码为新加上去的 -->
    <servlet>
        <servlet-name>springmvc</servlet-name>
        <servlet-class>org.springframework.web.servlet.DispatcherServlet</servlet-class>
        <init-param>
            <param-name>contextConfigLocation</param-name>
            <param-value>classpath:springmvc.xml</param-value>
        </init-param>
    </servlet>
    <servlet-mapping>
        <servlet-name>springmvc</servlet-name>
        <url-pattern>*.do</url-pattern>
    </servlet-mapping>
    <welcome-file-list>
      <welcome-file>index.html</welcome-file>
      <welcome-file>index.htm</welcome-file>
      <welcome-file>index.jsp</welcome-file>
      <welcome-file>default.html</welcome-file>
      <welcome-file>default.htm</welcome-file>
      <welcome-file>default.jsp</welcome-file>
    </welcome-file-list>
</web-app>
```

（这是默认原有代码，用于访问首页）

新添加的代码的意思是，凡是客户端发出的以".do"结尾的 URL 请求都会被 DispatcherServlet

核心控制器(又称中央调度器)拦截，DispatcherServlet 再交给 springmvc.xml 进行处理，springmvc.xml 接下来会为客户端的 url 请求找到对应的控制器 Controller 类进行处理。如果配置改为 "<url-pattern>/</url-pattern>"，则表示拦截所有请求。

（4）编写一个 Controller 类。

在包 com.lifeng.controller 下新建一个类 FirstController，实现 Controller 接口，代码如下：

```
package com.lifeng.controller;
import javax.servlet.http.HttpServletRequest;
import javax.servlet.http.HttpServletResponse;
import org.springframework.web.servlet.ModelAndView;
import org.springframework.web.servlet.mvc.Controller;
public class FirstController implements Controller{
    @Override
    public ModelAndView handleRequest(HttpServletRequest arg0, HttpServletResponse arg1) throws Exception {
        ModelAndView mv=new ModelAndView();
        mv.addObject("hello", "Hello SpringMVC!!!");
        mv.setViewName("welcome");
        return mv;
    }
}
```

这个方法的作用是创建一个 ModelAndView 对象，在该对象中添加一个名叫 "hello" 的字符串对象，值为 "Hello SpringMVC!!!"，并在 ModelAndView 对象中设置返回的逻辑视图为 welcome，这里逻辑视图 welcome 本身还不是完整的视图路径，将在下述有关步骤（视图解析器）进一步解释为完整路径。此方法意味着程序将携带 hello 对象数据跳转到名为 welcome 的逻辑视图。

（5）配置 handlerMapping 处理器映射器。在 src 下新建一个 XML 文件，将其命名为 springmvc.xml，输入如下内容：

```
<?xml version="1.0" encoding="UTF-8"?>
<beans xmlns="http://www.springframework.org/schema/beans"
    xmlns:xsi=http://www.w3.org/2001/XMLSchema-instance
    xmlns:mvc="http://www.springframework.org/schema/mvc"
    xmlns:context="http://www.springframework.org/schema/context"
    xmlns:aop=http://www.springframework.org/schema/aop
    xmlns:tx="http://www.springframework.org/schema/tx"
    xsi:schemaLocation="http://www.springframework.org/schema/beans
        http://www.springframework.org/schema/beans/spring-beans.xsd
        http://www.springframework.org/schema/mvc
        http://www.springframework.org/schema/mvc/spring-mvc.xsd
        http://www.springframework.org/schema/context
        http://www.springframework.org/schema/context/spring-context.xsd
        http://www.springframework.org/schema/aop
        http://www.springframework.org/schema/aop/spring-aop.xsd
        http://www.springframework.org/schema/tx
        http://www.springframework.org/schema/tx/spring-tx.xsd">
    <!-- 配置处理器映射器-->
    <bean class="org.springframework.web.servlet.handler.BeanNameUrlHandlerMapping">
    </bean>
</beans>
```

上述方框部分的代码意思是，创建一种类型为 BeanNameUrlHandlerMapping 的处理器映射器，即定义一种"请求/响应"映射规则，客户端的 URL 请求如果跟某一个 Bean 的 name 属性匹配，则由该 Bean 的 class 属性指定的控制器 Controller 类进行响应处理。

（6）配置处理器适配器。配置完处理器映射器后，接着在 springmvc.xml 中插入如下内容（插入位置在上图的方框下方，节点</beans>的上方）：

```
<bean class="org.springframework.web.servlet.mvc.SimpleControllerHandlerAdapter">
</bean>
```

该代码的意思是创建一种处理器适配器，类型为 SimpleControllerHandlerAdapter，用于对上述指定的控制器 Controller 类的 handleRequest()方法的调用与执行。

（7）配置自定义控制器。继续在 springmvc.xml 中插入如下内容：

```
<bean name="/hello.do" class="com.lifeng.controller.FirstController"></bean>
```

该代码自定义了一个 Bean，定义了一种具体的"请求/响应"映射关系，表示假如客户端发出的 URL 请求的是/hello.do，则指定由服务端的 com.lifeng.controller.FirstController 程序来处理，通过这行代码确定了一条具体的"请求/响应"映射关系，是一对一的对应关系，即 name 属性表示的客户端请求（如 URL 路径），对应 class 属性表示的服务端的响应程序。

【注意】这个 Bean 的配置要有效，前提是第 5 步要配置类型为 BeanNameUrlHandlerMapping 的映射器，以及第 6 步的适配器。第 5 步相当于制定了一个规则，如乒乓球比赛的"男女混合双打"，而第 7 步则是这个规则下的具体某某男与某某女进行配对混合双打。

（8）配置视图解析器。视图解释器用来解释控制器返回的逻辑视图的真实路径，这样更方便，易于扩展。在 springmvc.xml 中输入代码：

```
<!-- 4.配置视图解析器 -->
    <bean class="org.springframework.web.servlet.view.InternalResourceViewResolver">
        <!--逻辑视图前缀-->
        <property name="prefix" value="/WEB-INF/jsp/"></property>
        <!--逻辑视图后缀，匹配模式：前缀+逻辑视图+后缀，形成完整路径名-->
        <property name="suffix" value=".jsp"></property>
    </bean>
```

上面代码的意思是控制器 Controller 返回的逻辑视图，需要加上前缀 "/WEB-INF/jsp/">和后缀 ".jsp"，最后拼接成完整的视图路径。如本例中，Controller 返回的视图为 "welcome"，视图解释器将为它加上前缀和后缀，最终构成完整路径为 "/WEB-INF/jsp/ welcome.jsp"。视图解释器不是非要不可，如果没有视图解释器，则 Controller 返回的视图必须打上完整路径的视图名称，Controller 类应修改为如下所示：

```
public class FirstController implements Controller{
    @Override
    public ModelAndView handleRequest(HttpServletRequest arg0, HttpServletResponse arg1)
throws Exception {
        ModelAndView mv=new ModelAndView();
        mv.addObject("hello", "Hello SpringMVC!!!");
        mv.setViewName("/WEB-INF/jsp/welcome.jsp");
        return mv;
    }
}
```

至此完整的 springmvc.xml 内容结构如下所示：

```
<?xml version="1.0" encoding="UTF-8"?>
<beans xmlns="http://www.springframework.org/schema/beans"
    xmlns:xsi="http://www.w3.org/2001/XMLSchema-instance"
    xmlns:mvc="http://www.springframework.org/schema/mvc"
```

```xml
    xmlns:context="http://www.springframework.org/schema/context"
    xmlns:aop="http://www.springframework.org/schema/aop"
    xmlns:tx="http://www.springframework.org/schema/tx"
    xsi:schemaLocation="http://www.springframework.org/schema/beans
        http://www.springframework.org/schema/beans/spring-beans.xsd
        http://www.springframework.org/schema/mvc
        http://www.springframework.org/schema/mvc/spring-mvc.xsd
        http://www.springframework.org/schema/context
        http://www.springframework.org/schema/context/spring-context.xsd
        http://www.springframework.org/schema/aop
        http://www.springframework.org/schema/aop/spring-aop.xsd
        http://www.springframework.org/schema/tx
        http://www.springframework.org/schema/tx/spring-tx.xsd">
    <!-- 1.配置处理器映射器,确定一种请求~响应的映射规则-->
    <bean class="org.springframework.web.servlet.handler.BeanNameUrlHandlerMapping"></bean>
    <!-- 2.配置处理器适配器,配置对处理器的 handleRequest()方法的调用 -->
    <bean class="org.springframework.web.servlet.mvc.SimpleControllerHandlerAdapter"></bean>
    <!-- 3.配置自定义控制器,一条具体的映射关系,name 对应客户端 URL 请求,class 对应服务端响应程序 -->
    <bean name="/hello.do" class="com.lifeng.controller.FirstController"></bean>
    <!-- 4.配置视图解析器 -->
    <bean class="org.springframework.web.servlet.view.InternalResourceViewResolver">
        <!--逻辑视图前缀-->
        <property name="prefix" value="/WEB-INF/jsp/"></property>
        <!--逻辑视图后缀，匹配模式：前缀+逻辑视图+后缀，形成完整路径名-->
        <property name="suffix" value=".jsp"></property>
    </bean>
</beans>
```

提示：上述类型的处理器映射器是默认的，同样上述类型的处理器适配器也是默认的，两者均可省略不要，所以 springmvc.xml 又可以简化为下列代码。

```xml
<?xml version="1.0" encoding="UTF-8"?>
<beans xmlns="http://www.springframework.org/schema/beans"
    xmlns:xsi="http://www.w3.org/2001/XMLSchema-instance"
    xmlns:mvc="http://www.springframework.org/schema/mvc"
    xmlns:context="http://www.springframework.org/schema/context"
    xmlns:aop="http://www.springframework.org/schema/aop"
    xmlns:tx="http://www.springframework.org/schema/tx"
    xsi:schemaLocation="http://www.springframework.org/schema/beans
        http://www.springframework.org/schema/beans/spring-beans.xsd
        http://www.springframework.org/schema/mvc
        http://www.springframework.org/schema/mvc/spring-mvc.xsd
        http://www.springframework.org/schema/context
        http://www.springframework.org/schema/context/spring-context.xsd
        http://www.springframework.org/schema/aop
        http://www.springframework.org/schema/aop/spring-aop.xsd
        http://www.springframework.org/schema/tx
        http://www.springframework.org/schema/tx/spring-tx.xsd">
    <!-- 1.配置自定义控制器,一条具体的映射关系,name 对应客户端 URL 请求,class 对应服务端响应程序 -->
    <bean name="/hello.do" class="com.lifeng.controller.FirstController"></bean>
    <!-- 2.配置视图解析器 -->
```

```
            <bean class="org.springframework.web.servlet.view.InternalResourceViewResolver">
                <!--逻辑视图前缀-->
                <property name="prefix" value="/WEB-INF/jsp/"></property>
                <!--逻辑视图后缀，匹配模式：前缀+逻辑视图+后缀，形成完整路径名-->
                <property name="suffix" value=".jsp"></property>
            </bean>
</beans>
```

（9）定义一个响应页面。在目录 WebContent/WEB-INF/jsp 下新建一个 JSP 文件 welcome.jsp，打开 welcome.jsp 文件，在<body></body>之间输入如下代码：

```
<h1>${hello }</h1>
```

最终完整的项目结构如图 9.5 所示：

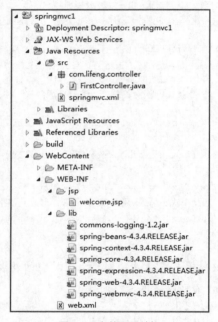

图 9.5 完整项目结构

（10）运行项目，在浏览器中输入 http://localhost:8080/springmvc1/hello.do，出现图 9.6 所示的页面。

图 9.6 运行结果

以上完整程序请见本书配套源码。

上机练习

在客户端浏览器的 URL 中输入 http://localhost:8080/mvc01/welcome.do，即可输出页面信息：欢迎光临。

思考题

1. 什么是 MVC 设计模式？
2. 简要说明 Spring MVC 的运行流程。
3. Spring MVC 需要哪些 JAR 包？

第 10 章　Spring MVC 注解式开发

本章目标

- ◆ 掌握注解式开发的基本步骤
- ◆ 掌握 DispatcherServlet 的配置方法
- ◆ 理解 Controller 注解
- ◆ 掌握 RequestMapping 注解方法
- ◆ 掌握参数的接收方式
- ◆ 理解 ModelAndView 返回类型
- ◆ 理解 String、void 返回类型
- ◆ 理解对象返回类型
- ◆ 掌握 Ajax 请求方式
- ◆ 掌握转发与重定向的方法

10.1 第一个注解式开发程序

第 9 章的第一个 Spring MVC 程序是基于配置式的开发，有助于我们理解 Spring MVC 的基本原理与流程，Spring MVC 还提供注解式的开发，大大简化了开发流程，实际开发中常使用注解式开发。下面来看第一个注解式开发程序。

项目案例：用注解式开发设计一个项目，在浏览器中输入 http://localhost:8080/springmvc2/first.do，输出网页内容"我的第一个注解式 Spring MVC"。（项目源码参见配套资源：第 10 章/第一个注解式 mvc 程序/springmvc2）

实现步骤：

（1）在 Eclipse 中新建 Dynamic Web Project 项目 springmvc2，导入有关 JAR 包，如图 10.1 所示。注意多了一个 spring-aop-4.3.4.RELEASE.jar 包，只有导入这个包才能使用注解。

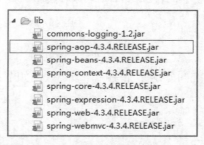

图 10.1 注解式开发需要导入的 JAR 包

（2）配置 web.xml。内容与第一个 Spring MVC 项目中的 web.xml 内容一模一样，复制过来即可。

（3）配置 springmvc.xml，需要配置组件扫描器，代码如下：

```xml
<?xml version="1.0" encoding="UTF-8"?>
<beans xmlns="http://www.springframework.org/schema/beans"
    xmlns:xsi="http://www.w3.org/2001/XMLSchema-instance"
    xmlns:mvc="http://www.springframework.org/schema/mvc"
    xmlns:context="http://www.springframework.org/schema/context"
    xmlns:aop="http://www.springframework.org/schema/aop"
    xmlns:tx="http://www.springframework.org/schema/tx"
    xsi:schemaLocation="http://www.springframework.org/schema/beans
        http://www.springframework.org/schema/beans/spring-beans.xsd
        http://www.springframework.org/schema/mvc
        http://www.springframework.org/schema/mvc/spring-mvc.xsd
        http://www.springframework.org/schema/context
        http://www.springframework.org/schema/context/spring-context.xsd
        http://www.springframework.org/schema/aop
        http://www.springframework.org/schema/aop/spring-aop.xsd
        http://www.springframework.org/schema/tx
        http://www.springframework.org/schema/tx/spring-tx.xsd">
<!-- 配置视图解释器 -->
<bean class="org.springframework.web.servlet.view.InternalResourceViewResolver">
    <!--逻辑视图前缀-->
    <property name="prefix" value="/WEB-INF/jsp/"></property>
    <!--逻辑视图后缀，匹配模式：前缀+逻辑视图+后缀，形成完整路径名-->
```

```xml
            <property name="suffix" value=".jsp"></property>
    </bean>
    <!-- 配置组件扫描器 -->
    <context:component-scan base-package="com.lifeng.controller"/>
</beans>
```

上面代码与配置式开发的相比，少了一大堆配置，无须配置处理器映射器和处理器适配器。但问题来了，那"请求/响应"的映射关系在哪定义呢？客户端的 URL 请求从哪里可以找到对应的服务端处理器来处理呢？可以发现上面代码里多了一个配置组件扫描器，表示通过扫描指定的包下面的类中的注解从而获取映射关系。这里是扫描包 com.lifeng.controller 下的所有类中的注解。

（4）创建控制器。在 src 下新建包 controller，新建一个类 FirstController。无须实现任何接口。初始代码如下：

```java
public class FirstController{
    public ModelAndView doHello() {
        ModelAndView mv=new ModelAndView();
        mv.addObject("msg", "我的第一个注解式SPringMVC!!!");
        mv.setViewName("welcome");
        return mv;
    }
}
```

接着在类名和方法 doHello()上面分别添加注解，最终代码如下：

```java
@Controller
public class FirstController{
    @RequestMapping("/hello.do")
    public ModelAndView doHello() {
        ModelAndView mv=new ModelAndView();
        mv.addObject("msg", "My First Annotation SpringMVC!!!");
        mv.setViewName("welcome");
        return mv;
    }
}
```

第一个注解@Controller 表示将本类定义为一个控制器类，这个类无须再实现 Controller 接口。

第二个注解@RequestMapping（"/hello.do"）表示定义一种"请求/响应"的映射关系，即如果客户端浏览器发出"/hello.do"的 URL 请求，则由该注解下面的 doHello()方法来响应，即浏览器通过 URL 路径+"/hello.do"就可访问到本方法，URL 请求能够直接映射到控制器类的方法级别。这样一个简单的注解，就轻松地取代了之前的处理器映射器和 Bean 的配置，大大减少了配置工作量。

（5）创建响应页面。

在项目 WebContent/WED-INF 下创建文件夹 jsp，再在目录 WebContent/WED-INF/jsp 下创建 welcome.jsp 页面。在<body></body>之间添加如下代码：

```
<h1>${msg}</h1>
```

（6）运行并测试程序。

在 URL 输入 http://localhost:8080/springmvc2/hello.do，输出网页内容：我的第一个注解式 SpringMVC。如果是中文可能会有乱码，后面会解决这个问题，这里先用英文，结果如图 10.2 所示。

图 10.2 注解式 MVC 输出页面

10.2 核心控制器 DispatcherServlet 的配置

核心控制器 DispatcherServlet 的典型配置如下：
```
<servlet>
      <servlet-name>springmvc</servlet-name>
      <servlet-class>org.springframework.web.servlet.DispatcherServlet</servlet-class>
      <init-param>
           <param-name>contextConfigLocation</param-name>
           <param-value>classpath:springmvc.xml</param-value>
      </init-param>
</servlet>
<servlet-mapping>
      <servlet-name>springmvc</servlet-name>
      <url-pattern>*.do</url-pattern>
</servlet-mapping>
```
它表示 Spring 将拦截所有的*.do 请求，其中，*可匹配任意字符，即拦截所有后缀为.do 的请求，再交给<init-param>下的<param-value>所指定的 springmvc.xml 文件进行解释。如果用/代替*.do，则是指拦截所有请求，代码如下所示：
```
<servlet-mapping>
      <servlet-name>springmvc</servlet-name>
      <url-pattern>/</url-pattern>
</servlet-mapping>
```
<param-value>classpath:springmvc.xml</param-value>指的是 Spring MVC 的配置文件为 src 目录下的 springmvc.xml 文件，容器启动后会自动到 src 目录下查找名为 springmvc.xml 的配置文件（然后加载进来）。

【注意】 如果 Spring MVC 找不到配置文件就会报错。

<init-param>节点是可选项，可以省略，则 web.xml 中的核心控制器 DispatcherServlet 的配置可简化为如下代码：
```
<servlet>
       <servlet-name>springmvc</servlet-name>
       <servlet-class>org.springframework.web.servlet.DispatcherServlet</servlet-class>
   </servlet>
   <servlet-mapping>
       <servlet-name>springmvc</servlet-name>
       <url-pattern>*.do</url-pattern>
   </servlet-mapping>
```
这里并未指明 Spring MVC 配置文件的位置，问题来了，Spring 拦截所有*.do 请求后，下一步交

给谁进行解释呢,即到哪里找到 Spring MVC 配置文件呢?答案是应用程序在启动时会默认到 WEB-INF 目录下查找如下格式的 MVC 配置文件:

ServletName-servlet.xml

其中 ServletName 应与 web.xml 配置文件的<servlet-name>中的一致。

对开发者来讲,这个 MVC 配置文件需自行创建,而且要遵守两条规则:

① 命名规则:ServletName-servlet.xml;

② 位置规则:必须放在 WEB-INF 下。

本例中由于 web.xml 中的 ServletName 为 springmvc,故 mvc 配置文件应命名为 springmvc-servlet.xml,且必须放在目录 WEB-INF 下。

此外核心控制器 DispatcherServlet 的配置还可选用<load-on-startup>1</load-on-startup>表示 Serlvet 容器在服务器启动时立即加载。如果没有这条配置,则表示应用程序在第一个 Servlet 请求时加载该 Servlet。

10.3　@Controller 注解

传统的配置式开发中的控制器 Controller 类必须实现 Controller 接口,并实现接口中的 HandleRequest()方法,还需要在配置文件中配置处理器映射,且一个处理器(控制器)只能有一个方法,为了实现程序功能,不得不创建大量的处理器(控制器)类,不够方便灵活。而一个基于注解的控制器可以有多个方法,能大大减少处理器(控制器)类的数量。

将一个普通的类转化为控制器类只需要在该类的声明上面加上@Controller 注解即可。要使@Controller 注解被扫描到,必须在 springmvc.xml 文件中配置组件扫描器,代码如下:

```
<!-- 配置组件扫描器 -->
<context:component-scan base-package="controller"/>
```

其中 base-package 指明使用了注解的 Controller 类所在的包,系统会自动扫描该包下的所有类,识别其中的注解。

10.4　@RequestMapping 注解

该注解可以用于类上,也可用于方法上。

10.4.1　注解用于方法上

基于注解的控制器无须在 xml 配置文件中配置处理器映射器,仅需要用@RequestMapping 对控制器类中任意一个方法进行注解即可建立"请求/响应"映射关系,将客户端请求与处理器的方法一一对应。

项目案例:客户端浏览器如果请求 login.do,则出现登录界面,如果请求 register.do,则出现注册界面。使用 Spring MVC 注解式开发。(项目源码参见配套资源:第 10 章/RequestMapping 注解用于方法上/springmvc3)

实现步骤：

（1）创建项目 springmvc3，添加 JAR 包，配置 web.xml，代码参照上一个项目。

（2）在 SRC 下创建 springmvc.xml 配置文件，代码参照上一个项目。

（3）创建控制器类 UserController，代码如下：

```java
package controller;
import org.springframework.stereotype.Controller;
import org.springframework.web.bind.annotation.RequestMapping;
@Controller
public class UserController {
    @RequestMapping("/login.do")
    public String login(){
        return "login";
    }
    @RequestMapping("/register.do")
    public String register(){
        return "register";
    }
}
```

可以看到，UserController 类下的 login() 和 register() 方法都添加了注解，分别表示客户端通过 URL 请求"/login.do"就可访问到 login() 方法，通过 URL 请求"/register.do"就可访问到 register() 方法。注意，login 方法在这里是 String 类型，并不是指返回字符串给客户端，而是指返回给客户端的逻辑视图名称，再经 sprimgmvc.xml 配置文件的中视图解释器进行解释，即拼接前缀与后缀，才返回实际的视图给客户端。

（4）在 WEB-INF/jsp 目录下创建 login.jsp 和 register.jsp 文件。

login.jsp 关键代码如下：

```
<body>
用户登录
<form action="">
姓名:<input type="text" name="name"/><br/>
密码:<input type="text" name="password"/><br/><br/>
<input type="submit" value="登录"/>
</form>
</body>
```

register.jsp 关键代码如下：

```
<body>
用户注册
<form action="">
姓名:<input type="text" name="username"/><br/>
密码:<input type="text" name="password"/><br/><br/>
<input type="submit" value="注册"/>
</form>
</body>
```

（5）运行测试，可以发现在浏览器中输入 http://localhost:8080/springmvc3/login.do，会出现图 10.3 所示的用户登录界面。

图 10.3　用户登录界面

在浏览器中输入 http://localhost:8080/springmvc3/register.do，会出现图 10.4 所示的用户注册界面。

图 10.4　用户注册界面

10.4.2　注解用于类上

在上面的项目中，客户端通过 URL 请求"/login.do"就可访问到 UserController 类的 login()方法，通过 URL 请求"/register.do"就可访问到 UserController 类的 register ()方法。

如果该项目只有用户登录，则没有什么冲突，但如果该项目除了用户登录，还有一种管理员登录，假定管理员登录注册时也用到同一个客户端 URL 请求路径。实现管理员登录注册的 Controller 类 AdminController 代码如下：

```
package controller;
import org.springframework.stereotype.Controller;
import org.springframework.web.bind.annotation.RequestMapping;
@Controller
public class AdminController {
    @RequestMapping("/login.do")
    public String login(){
        return "loginAdmin";
    }
    @RequestMapping("/register.do")
    public String register(){
        return "registerAdmin";
    }
}
```

可以看到，控制器中 AdminController 的两个@RequestMapping 注解都与控制器 UserController 的@RequestMapping 注解相同。问题来了，若浏览器发出/login.do 的 url 请求，那到底该返回用户登录界面，还是管理员登录界面呢？

现在的情况是无法区分，程序将报错，因为两个类的 login()方法用的都是同一种注解，出现冲突。假定两个类的 login()方法的注解都不变，有什么办法可以解决冲突问题呢？

解决办法是在各自的类上再加上一个@RequestMapping 注解，且每个类的注解的名称不同。具体操作见下面这个案例。

项目案例：把项目 springmvc3 复制为 springmvc4，增加管理员登录和注册功能，解决上述冲突问题。（项目源码参见配套资源：第 10 章/RequestMapping 注解用于类上/springmvc4）

实现步骤：

（1）把项目 springmvc3 复制为 springmvc4，添加 AdminController 类代码同上。注意 Web 项目复制后需要用鼠标右键单击项目，选择 properties，修改属性中的 Web Project Settings→Context root。

（2）将 login.jsp 页面复制为 loginAdmin.jsp，内容修改如下：

```
<%@ page language="java" contentType="text/html; charset=utf-8"
    pageEncoding="utf-8"%>
<!DOCTYPE html PUBLIC "-//W3C//DTD HTML 4.01 Transitional//EN" "http://www.w3.org/TR/html4/loose.dtd">
<html>
<head>
<meta http-equiv="Content-Type" content="text/html; charset=utf-8">
<title>Insert title here</title>
</head>
 <body>
管理员登录
<form action="">
姓名:<input type="text" name="name"/><br/>
密码:<input type="text" name="password"/><br/><br/>
<input type="submit" value="登录"/>
</form>
</body>
</html>
```

（3）将 register.jsp 页面复制为 registerAdmin.jsp，内容修改如下：

```
<%@ page language="java" contentType="text/html; charset=utf-8"
    pageEncoding="utf-8"%>
<!DOCTYPE html PUBLIC "-//W3C//DTD HTML 4.01 Transitional//EN" "http://www.w3.org/TR/html4/loose.dtd">
<html>
<head>
<meta http-equiv="Content-Type" content="text/html; charset=utf-8">
<title>Insert title here</title>
</head>
<body>
管理员注册
<form action="">
姓名:<input type="text" name="username"/><br/>
密码:<input type="text" name="password"/><br/><br/>
<input type="submit" value="注册"/>
</form>
</body>
</html>
```

（4）测试运行。在浏览器中输入 http://localhost:8080/springmvc4/login.do，程序报错。原因刚才已经提到了，就是两个控制类上的注解都是/login.do。

（5）解决方案是分别给两个控制类上添加@RequestMapping 注解。

本例中 UserController 类添加@RequestMapping 注解后代码如下：
```
package com.lifeng.controller;
import org.springframework.stereotype.Controller;
import org.springframework.web.bind.annotation.RequestMapping;
@Controller
@RequestMapping("/user")
public class UserController {
    @RequestMapping("/login.do")
    public String login(){
        return "login";
    }
    @RequestMapping("/register.do")
    public String register(){
        return "register";
    }
}
```

AdminController 类添加@RequestMapping 注解后代码如下：
```
package controller;
import org.springframework.stereotype.Controller;
import org.springframework.web.bind.annotation.RequestMapping;
@Controller
@RequestMapping("/admin")
public class AdminController {
    @RequestMapping("/login.do")
    public String login(){
        return "login";
    }
    @RequestMapping("/register.do")
    public String register(){
        return "register";
    }
}
```

除此之外，URL 请求路径还要加上类上的注解作为上一级路径，这样两种不同的登录就可以区分开了。

如果访问 http://localhost:8080/springmvc4/user/login.do，则访问到的是图 10.5 所示的用户登录界面。

图 10.5　访问到用户登录界面

如果访问 http://localhost:8080/springmvc4/admin/login.do，则访问到的是图 10.6 所示的管理员登录界面。

图 10.6　访问到管理员登录界面

结论：当类和方法上都有@RequestMapping 注解时，最终的 URL 请求必须是两者的结合体。例如，类上的@RequestMapping 注解为@RequestMapping("/user")，方法上的@RequestMapping 注解为@RequestMapping("/login.do")，则最终的 URL 请求为/user/login.do，单独用方法上注解的/login.do 请求将无法被识别，必须先加上类上的注解/user 才能构成完整的请求路径。类上的注解是上一级路径，方法上的注解是下一级路径。

10.4.3　请求的提交方式

在项目 springmvc3 中，若要限定只能用 GET 方式访问 login()方法，则需要将 login()方法上的@RequestMapping 注解修改如下：

```
@RequestMapping(value="/login.do",method=RequestMethod.GET)
public String login(){
    return "login";
}
```

同理，若要限定只能用 POST 方式访问 login()方法，则需要将 login()方法上的@RequestMapping 注解修改如下：

```
@RequestMapping(value="/login.do",method=RequestMethod.POST)
public String login(){
    return "login";
}
```

这时若强行用 GET 方式访问，将出现图 10.7 所示的错误。

图 10.7　提交方式报错

@RequestMapping 在默认情况下两种方式都支持，如果不想限制 GET 或 POST 方式，则删除@RequestMapping 注解中的 method=RequestMethod.GET 或 method=RequestMethod.POST 即可。

项目案例：将项目 springmvc3 复制为 springmvc5，做上述修改，测试效果。（项目源码参见配套资源：第 10 章/限定 post 或 get 提交方式/springmvc5）

关键步骤：

（1）将 UserController 类修改如下：

```
package com.lifeng.controller;
import org.springframework.stereotype.Controller;
import org.springframework.web.bind.annotation.RequestMapping;
import org.springframework.web.bind.annotation.RequestMethod;
@Controller
public class UserController {
    //限定为 GET 提交方式
    //@RequestMapping(value="/login.do",method=RequestMethod.GET)
    //限定为 POST 提交方式
    @RequestMapping(value="/login.do",method=RequestMethod.POST)
    public String login(){
        return "login";
    }
    @RequestMapping("/register.do")
    public String register(){
        return "register";
    }
}
```

（2）分别用两种方式进行测试。

10.4.4　请求 URI 中使用通配符

在资源路径中使用通配符，可以匹配灵活多变的 URI 请求，映射到同一个控制器方法进行响应，有两种用法：路径级数的精确匹配、路径级数的可变匹配。

/user/*/login.do：表示在 login.do 的资源名称前面，限定只能有两级路径，第一级路径必须是/user，而第二级路径随意。这称为路径级数的精确匹配。

/user/**/login.do：表示在 login.do 的资源名称前面，必须以/user 路径开头，而其他级的路径可有可无。若有，又包含几级，各级又叫什么名称，都随意。这称为路径级数的可变匹配。

项目案例：将项目 springmvc4 复制为 springmvc6，分别测试路径级数的精确匹配和可变匹配。（项目源码参见配套资源：第 10 章/路径级数的精确匹配与模糊匹配/springmvc6）

关键步骤：

（1）将项目 springmvc4 复制为 springmvc6，修改 UserController 如下：

```
package com.lifeng.controller;
import org.springframework.stereotype.Controller;
import org.springframework.web.bind.annotation.RequestMapping;
@Controller
 @RequestMapping("/user")
public class UserController {
    @RequestMapping("/*/login.do")      ← 路径级数精确匹配
    public String login(){
        return "login";
    }
    @RequestMapping("/**/register.do")  ← 路径级数可变匹配
    public String register(){
```

```
            return "register";
    }
}
```

（2）测试。在浏览器中输入 http://localhost:8080/springmvc6/user/login.do，报错，级数不匹配。
再次输入 http://localhost:8080/springmvc6/user/aa/login.do，正确显示，如图10.8所示。

图 10.8　路径级数精确匹配

再次输入如下地址：

http://localhost:8080/springmvc6/user/aa/bb/register.do

http://localhost:8080/springmvc6/user/aa/register.do

http://localhost:8080/springmvc6/user/register.do

均可正常显示，如图10.9所示。

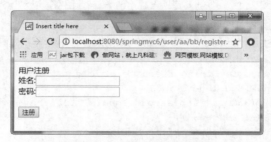

图 10.9　路径级数可变匹配

10.4.5　请求中携带参数

@RequestMapping 中 params 属性定义了请求中必须携带的参数的要求。以下是几种情况的说明。

（1）@RequestMapping(value="/login.do",params={"name","pwd"})：要求请求中必须携带请求参数 name 与 pwd。

（2）@RequestMapping(value="/login.do",params={"!name","pwd"})：要求请求中必须携带请求参数 pwd，但必须不能携带参数 name。

（3）@RequestMapping(value="/login.do",params={"name=john","pwd=123456"})：要求请求中必须携带请求参数 name，且其值必须为 john；必须携带参数 pwd，且其值必须为 123456。

（4）@RequestMapping(value="/login.do",params="name!= john")：要求请求中必须携带请求参数 name，且其值必须不能为 john。

@RequestMapping 注解的各个属性，见表 10.1。

表 10.1 RequestMapping 注解的各个属性

属性名称	说明
value	默认属性，映射一个客户端请求
method	用于限定提交方式，POST 或 GET，默认两者都支持
params	参数，限定请求中必须包含或不包含哪些参数
headers	限定请求中必须包含某些指定的 header 的值
consumes	指定请求提交的内容类型，常用的类型有 application/json 和 ext/html 等
produces	指定返回的内容类型，常用的类型有 application/json 和 text/html 等

10.5 客户端到处理器的参数传递

10.5.1 基本类型做形式参数

要求前台页面的表单输入框的 name 属性与对应控制器方法中的形式参数名称与类型一致，控制器方法就能接收到来自前台表单传过来的参数，即请求参数与方法形参要完全相同，这些参数由系统在调用时直接赋值，程序员可在方法内直接使用。

项目案例： 如果用户输入有用户名为 admin，密码为 123，则返回登录成功的界面，否则返回登录错误的页面。（项目源码参见配套资源：第 10 章/通过方法的形参接收客户端数据/springmvc7）

关键步骤：

（1）复制 springmvc4 为 springmvc7，在 UserController 类中新建一个方法，代码如下所示：

```
@RequestMapping("/dologin.do")
public ModelAndView doLogin(String username,String password){
    ModelAndView mv=new ModelAndView();
    if(username.equals("admin")&&password.equals("123")){
        mv.setViewName("success");
        mv.addObject("user","admin");
    }else{
        mv.setViewName("fail");
    }
    return mv;
}
```

（2）login.jsp 页面修改如下：

```
<body>
用户登录
<form action="dologin.do">
姓名:<input type="text" name="username"/><br/>
密码:<input type="text" name="password"/><br/>
<input type="submit" value="登录"/>
</form>
</body>
```

【注意】 添加了 action="dologin.do"。

（3）新建 success.jsp 和 fail.jsp 页面，简单地分别加一句登录成功、登录失败即可测试成功失败两种情形。

success.jsp 关键代码如下：
`<h1>登录成功,欢迎${user }光临</h1>`

fail.jsp 关键代码如下：
`<h1>登录失败</h1>`

（4）在浏览器中输入 http://localhost:8080/springmvc6/user/login.do 测试成功和失败的登录，如图 10.10 所示。

图 10.10　登录成功

结论：前台表单的数据提交后被后台的 dologin()方法成功获取，证明了只要前台表单的 name 属性和方法的形式参数名称与类型一致即可传递或获取数据。

【注意】如果表单的 name 属性与方法的参数名称不一致，也有办法，就是用@RequestParam()校正参数名，做法是在接收方法的形参前加个@RequestParam("表单的 name 属性")，示例如下。

假定表单的 name 属性分别为 username 和 password，而处理器的方法参数分别为 uname 和 pwd，代码如下：

```
@RequestMapping("/dologin.do")
public ModelAndView doLogin(@RequestParam("username") String uname, @RequestParam
("password") String pwd){
    ModelAndView mv=new ModelAndView();
    if(uname.equals("admin")&&pwd.equals("123")){
        mv.addObject("user","admin");
        mv.setViewName("admin");
    }else{
        mv.setViewName("fail");
    }
    return mv;
}
```

上面代码中，方法的形参分别是 uname、pwd，明显与表单不一样，但形参 uname 前面加上了注解@RequestParam("username")，表示先接收到表单 name 属性为 username 的值，再传给 uname。同理，形参 pwd 前面加上了注解@RequestParam("password")，表示先接收到表单 name 属性为 password 的值，再传给 pwd。

@RequestParam()有 3 个属性。

① value：指定请求参数的名称。

② required：指定该注解所修饰的参数是否是必须的，boolean 类型。若为 true，则表示请求中所携带的参数必须包含当前参数。若为 false，则表示有没有均可。

③ defaultValue：指定当前参数的默认值。若请求 URI 中没有给出当前参数，则当前方法参数将

取该默认值。即使 required 为 true，且 URI 中没有给出当前参数，该处理器方法参数会自动取该默认值，而不会报错。

10.5.2 中文乱码问题

上面案例所请求的参数若含有中文，可能会出现中文乱码问题，Spring MVC 对请求参数中的中文乱码问题，提供了专门的字符集过滤器，在 web.xml 配置文件中注册字符串过滤器即可解决中文乱码问题。上面项目若要解决乱码问题，在 web.xml 中添加如下配置即可。（项目源码参见配套资源：第 10 章/通过方法的形参接收客户端数据/springmvc7）

```xml
<!--注册字符集过滤器-->
<filter>
        <filter-name>characterEncodingFilter</filter-name>
        <filter-class>org.springframework.web.filter.CharacterEncodingFilter</filter-class>
        <init-param>
<!--指定字符集-->
            <param-name>encoding</param-name>
            <param-value>UTF-8</param-value>
        </init-param>
        <init-param>
<!--强制使用指定字符集-->
            <param-name>forceEncoding</param-name>
            <param-value>true</param-value>
        </init-param>
    </filter>
    <filter-mapping>
        <filter-name>characterEncodingFilter</filter-name>
        <url-pattern>/*</url-pattern>
    </filter-mapping>
```

运行该项目，测试用中文登录。

10.5.3 实体 Bean 做形参

项目 springmvc7 中，方法 doLogin()也可只用一个实体类作形式参数，前提是这个实体类的各个属性要与前台表单的各个 name 属性相同。

项目案例：如果用户输入有用户名为 admin，密码为 123，则返回登录成功的界面，否则返回登录错误的页面。（项目源码参见配套资源：第 10 章/实体 bean 整体接收客户端数据/springmvc8）

关键步骤：

（1）复制项目 springmvc7 为 springmvc8，新建包 com.lifeng.entity，包下新建一个实体类 User，代码如下：

```java
package com.lifeng.entity;
public class User {
    String username;
    String password;
    //省略getter,setter方法
}
```

这里的两个属性名称均要与前台表单的 name 属性一致。

（2）修改 dologin()方法如下：
```
@RequestMapping("/dologin.do")
public String doLogin(User user){
    if(user.getUsername().equals("admin")
            &&user.getPassword().equals("123")){
        System.out.println(user.getUsername());
        System.out.println(user.getPassword());
        return "success";
    }
    return "login";
}
```
运行测试后发现一样能登录，并且控制台有输出，证明实体 Bean 成功接收到了参数，如果实体 Bean 下的某一个属性又是对象类型会怎么样，还能接收到请求参数吗？

10.5.4 实体 Bean 含对象属性

当请求参数为某类对象属性的属性值时，要求请求参数名称为"类对象属性.属性"。

项目案例：在项目 springmvc8 实现了简单注册功能的基础上，若注册时还要提供地址信息，而地址信息封装在一个名为 Address 的 Bean 类里面，包含国家和城市属性，User 下面添加了一个类型为 Address 的对象属性，如何获得包含地址信息在内的注册信息？（项目源码参见配套资源：第 10 章/实体 bean 包含对象属性/springmvc9）

实现步骤：

（1）复制 springmvc8 为 springmvc9，在包 com.lifeng.entity 下新建 Address 类，代码如下：
```
package com.lifeng.entity;
public class Address {
    String country;
    String city;
//省略 getter,setter 方法
}
```
（2）修改 User 类，添加 Address 类型的属性，代码如下：
```
package com.lifeng.entity;
public class User {
    String username;
    String password;
    Address address;
    //省略 getter,setter 方法
}
```
（3）修改 register.jsp 页面，增加地址方面的注册信息，代码如下：
```
<body>
用户注册
<form action="doregister.do">
姓名:<input type="text" name="name"/><br/>
密码:<input type="text" name="password"/><br/>
国家:<input type="text" name="address.country"/><br/>
城市:<input type="text" name="address.city"/><br/>
<input type="submit" value="注册"/>
```

</form>
</body>

（4）在控制器中新增 doRegister()方法如下：

```
@RequestMapping("/doregister.do")
public ModelAndView doRegister(User user){
    ModelAndView mv=new ModelAndView();
    mv.setViewName("regsuccess");
    mv.addObject("user",user);
    System.out.println(user.getAddress().getCountry());//测试是否接收到国家
    System.out.println(user.getAddress().getCity());//测试是否接收到城市
    return mv;
}
```

（5）在 WebContent/WEB-INF/jsp 目录下新建 regsuccess.jsp 页面，代码如下：

```
<body>
用户注册信息
<form action="doregister.do">
用户名:${user.username}<br/>
密码:${user.password}<br/>
国家:${user.address.country}<br/>
城市:${user.address.city}<br/>
</body>
```

（6）运行测试，可以发现国家和城市信息成功地从前台传给了后台，并再次传回前台，如图 10.11 和图 10.12 所示。

图 10.11　注册界面

图 10.12　注册成功界面

10.5.5　路径变量

方法的形参，除了可以接收请求中携带的参数，也可接收 URL 中的变量，例如 URL 路径 http://localhost:8080/springmvc10/zhangsan/1/user.do 里面含有 zhangsan，显然对应的是用户名的信息，里面含有 1，对应的是用户编号 id，这些信息通过@RequestMapping 注解配合方法参数名称，一样能接收到。

项目案例：接收 URL 路径变量。（项目源码参见配套资源：第 10 章/路径变量的接收/springmvc10）

实现步骤：

（1）复制项目 springmvc9 为 springmvc10，在控制器 UserController 中添加方法 doUser()，代码如下：

```
@RequestMapping("/{username}/{id}/user.do")
public ModelAndView doUser(@PathVariable String username, @PathVariable String id){
    ModelAndView mv=new ModelAndView();
    mv.setViewName("success");
    System.out.println(username);//测试是否接收到 username
```

```
        System.out.println(id);//测试是否接收到id
        return mv;
    }
```
以上代码中，"/{username}/{id}/user.do"用{}括起来的部分被称为路径变量，将指定的URL模式地址与访问控制器指定方法的路径进行关联和匹配，如果可匹配，则控制器的指定方法被调用，并且路径中{}部分会被当作变量值被方法接收。

一般情况下，路径变量名要与方法中的形参名称一致。例如，@RequestMapping中的{username}要与方法doUser(String username,String id)中的username相同。

方法的每一个参数前面都要用注解@PathVariable修饰，否则接收不到数据。

（2）测试http://localhost:8080/springmvc10/user/zhangsan/1/user.do，观察控制台的输出。

（3）如果路径变量与方法参数名称不一致，则需用到@PathVariable进行修正，上面方法修改如下：
```
@RequestMapping("/{uname}/{uid}/user.do")
    Public ModelAndView doUser(@PathVariable("uname")String username,@PathVariable("uid")String id){
        ModelAndView mv=new ModelAndView();
        mv.setViewName("success");
        System.out.println(username);//测试是否接收到username
        System.out.println(id);//测试是否接收到id
        return mv;
    }
```

10.5.6　RESTful风格编程

RESTful风格是把请求参数变为请求路径的一种编程风格。通过路径变量的使用，可以实现RESTful风格的编程。传统的编程风格中，某项事务列表Web页面，要想一个个编辑，需要每一项中有如下这种超链接：

`/detail?id=1`

其中每一项的id均不同。而采用RESTful风格后，超链接将改为：

`/detail/1`

或者

`1/detail`

项目案例：学生列表页面超链接了学生详细信息页面，详细步骤如下，先用传统风格编程，再改用RESTful风格。（项目源码参见配套资源：第10章/REST风格/springmvc11）

实现步骤：

（1）将项目springmvc10复制为springmvc11，首先修改web.xml，将<url-pattern>/*.do</url-pattern>修改为<url-pattern>/</url-pattern>，表示拦截所有请求，然后在WebContent/WEB-INF/jsp下添加页面list.jsp，关键代码如下：
```
<body>
<h2 style="text-align:center">学生列表</h2>
<table>
<tr>
<td>编号:</td><td>姓名:</td><td>性别:</td><td>年龄:</td>
<td>详细信息 </td>
</tr>
<tr>
```

```
<td>1</td><td>张无忌</td><td>男</td><td>22</td>
<td><a href="detail?id=1">详细信息 </a></td>
</tr>
<tr>
<td>2</td><td>李寻欢</td><td>男</td><td>24</td>
<td><a href="detail?id=2">详细信息 </a></td>
</tr>
<tr>
<td>3</td><td>赵敏</td><td>女</td><td>19</td>
<td><a href="detail?id=3">详细信息 </a></td>
</tr>
</table>
</body>
```

再添加 detail.jsp 页面如下：

```
<%@ page language="java" contentType="text/html; charset=utf-8"
    pageEncoding="utf-8"%>
<!DOCTYPEhtmlPUBLIC"-//W3C//DTDHTML4.01Transitional//EN" "http://www.w3.org/TR/html4/loose.dtd">
<html>
<head>
<meta http-equiv="Content-Type" content="text/html; charset=ISO-8859-1">
<title>Insert title here</title>
</head>
<body>
<h1>你正在浏览的学生的学号是${id }</h1>
</body>
</html>
```

（2）在控制器 UserController 中添加如下两个方法：

```
@RequestMapping("/list ")
public String list(){
    return "list";
}
@RequestMapping("/detail")
public ModelAndView detail(int id){
    ModelAndView mv=new ModelAndView();
    mv.addObject("id", id);
    mv.setViewName("detail");
    return mv;
}
```

（3）运行测试，输入 http://localhost:8080/user/springmvc11/list，出现下面页面。把光标放在其中一个超链接上，状态栏显示其超链接是传统的 detail?id=3 形式，如图 10.13 所示。

图 10.13　学生列表

再单击超链接,即会出现图 10.14 所示的页面。

图 10.14　学生详细页面

(4)下面改用 RESTful 风格编程,复制 list.jsp 为 list2.jsp,修改代码如下:
```
<body>
<h2 style="text-align:center">学生列表</h2>
<table>
<tr>
<td>编号:</td><td>姓名:</td><td>性别:</td><td>年龄:</td>
<td>详细信息 </td>
</tr>
<tr>
<td>1</td><td>张无忌</td><td>男</td><td>22</td>
<td><a href="details/1">详细信息 </a></td>
</tr>
<tr>
<td>2</td><td>李寻欢</td><td>男</td><td>24</td>
<td><a href="details/2">详细信息 </a></td>
</tr>
<tr>
<td>3</td><td>赵敏</td><td>女</td><td>19</td>
<td><a href="details/3">详细信息 </a></td>
</tr>
</table>
</body>
</html>
```
(5)在控制器 UserController 中添加如下两个方法:
```
@RequestMapping("/list2 ")
public String list2(){
    return "list2";
}
@RequestMapping("/details/{id}")
public ModelAndView details(@PathVariable int id){
    ModelAndView mv=new ModelAndView();
    mv.addObject("id", id);
    mv.setViewName("detail");
    return mv;
}
```
(6)运行测试,输入 http://localhost:8080/springmvc11/user/list2,将会出现图 10.15 所示的页面。注意,将光标移到最后一行的详细信息上面时,状态栏显示的超链接显然为 RESTful 风格。

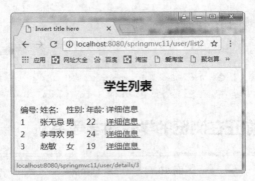

图 10.15　学生列表

再次单击超链接，会出现图 10.16 所示的页面，注意 URL 栏的不同。

图 10.16　详细页面

这样就实现了 RESTful 风格的编程。

10.5.7　HttpServletRequest 参数

使用 HttpServletRequest 对象、HttpSession 对象均可接收到请求参数。下面以 HttpServletRequest 为例进行说明。

项目案例：HttpServletRequest 可接收请求参数。doLogin()方法修改如下，同样可以接收到请求参数，实现登录，这种方法类似传统的 JSP+Servlet 的方式。（项目源码参见配套资源：第 10 章/HttpServletRequest 接收请求参数/springmvc1102）

关键步骤：

（1）将项目 springmvc10 复制为 springmvc1102，修改控制器 UserController 的 doLogin 方法如下：

```
@RequestMapping("/dologin.do")
public ModelAndView doLogin(HttpServletRequest request){
    String username=request.getParameter("username");
    String password=request.getParameter("password");
    ModelAndView mv=new ModelAndView();
    if(username.equals("张三")&&password.equals("123")){
        mv.setViewName("success");
        mv.addObject("user","张三");
    }else{
        mv.setViewName("fail");
    }
    return mv;
}
```

（2）将 WEB-INF/jsp 下的 success.jsp 页面修改如下：
```
<body>
<h1>登录成功,欢迎${user}光临</h1>
</body>
```
（3）运行测试。

10.5.8　接收数组类型的请求参数

控制器的方法除了接收基本类型的请求参数，也能整体接收 Bean 类型的参数，还能接收数组参数。接收数组参数有两个关键点：①前台表单有多个表单域的 name 属性相同；②控制器方法用这个 name 值命名的数组作为参数。

项目案例：页面有多个兴趣爱好供选择，选择好后，控制台能显示出来。（项目源码参见配套资源：第 10 章/接收数组类型的请求参数/springmvc1103）

关键步骤：

（1）复制项目 springmvc1102 为 springmvc1103，修改 UserController，添加下面两个方法：
```
@RequestMapping("/interest.do")
    public String interest(){
        return "interest";
    }
    @RequestMapping("/dointerest.do")
    public String doInterest(String[] myinterest){
        //测试用控制台输出各条兴趣
        System.out.println("我的兴趣爱好有:");
        for(String interest:myinterest){
            System.out.print(interest+" ");
        }
        return "interest";
    }
```

> 同名

（2）在 WEB-INF/jsp 下新增页面 interest.jsp，代码如下：
```
<body>
我的兴趣爱好<br/>
<form action="dointerest.do">
摄影:<input type="checkbox" name="myinterest" value="摄影"/><br/>
跳舞:<input type="checkbox" name="myinterest" value="跳舞"/><br/>
旅游:<input type="checkbox" name="myinterest" value="旅游"/><br/>
阅读:<input type="checkbox" name="myinterest" value="阅读"/><br/>
<input type="submit" value="确定"/>
</form>
<br/>观测控制台的输出
</body>
```
（3）运行测试，输入 http://localhost:8080/springmvc1103/user/interest.do 测试控制台输出。

10.6　服务端到客户端的参数传递

上面学习了参数从客户端到服务端的传递，下面来学习参数从服务端到客户端的传递。使用

@Controller 注解的控制器下面的方法，其返回值有以下 4 种类型：ModelAndView、String、无返回值 void、返回自定义类型对象。

若控制器方法处理完后，需要跳转到其他资源，同时又要往跳转的资源传递数据，此时控制器方法返回 ModelAndView 比较合适。这种情况下在控制器方法里面需要定义一个 ModelAndView 对象，并调用 ModelAndView 对象的 addObject()方法添加 Model 数据，用于传递到目标 View 页面，调用 setViewName()方法设置目标 View 页面。这在上面的案例中已多次用到。

若该控制器方法只跳转，而不传数据，或只传数据，而不向任何资源跳转（如页面的 Ajax 异步响应），此时若返回 ModelAndView，都不太合适，要么 Model 多余，要么 View 多余。可以考虑用其他 3 种返回值类型。

除了 ModelAndView 可以把数据从后台带到前端，还有 HttpServletRequest、Model 和 HttpSession 等对象也可以，后面章节将会介绍。

10.7 控制器方法返回 String 类型

控制器方法返回的字符串代表的是逻辑视图名，再通过 InternalResourceViewResolver 内部资源视图解析器解析将其转换为物理视图地址。除了逻辑视图名，也可以是 View 对象名，但需要另外定义一个 BeanNameViewResolver 视图解释器将其解释为真正的 URL。

1. 返回内部资源逻辑视图名

若要跳转的资源为内部资源，则视图解析器可以使用 InternalResourceViewResolver 内部资源视图解析器。此时处理器方法返回的字符串就是要跳转页面的文件名去掉前缀路径和后缀文件扩展名后的部分。这个字符串与视图解析器中的 prefix、suffix 相结合，即可形成要访问的 URI。之前的案例大都属于这种情况。

2. 返回 View 对象名

若要跳转的资源为外部资源，则可以使用 BeanNameViewResolver 类型的视图解析器，然后在配置文件中再定义一些外部资源视图 View 对象，此时处理器方法返回的字符串就是要跳转资源视图 View 的名称。当然，这些视图 View 对象，也可以是内部资源视图 View 对象。一般这种情况只跳转，不带传递数据。

10.7.1 返回 View 对象名

项目案例：将项目 springmvc11 复制为 springmvc12，实现登录成功后跳转到百度首页，否则跳转到 163 网易邮箱页面。（项目源码参见配套资源：第 10 章/返回字符串跳转到外部 URI/springmvc12）

实现步骤：

（1）将项目 springmvc11 复制为 springmvc12，修改控制器方法 doLogin()如下：

```
@RequestMapping("/dologin.do")
    public String doLogin(User user){
        if(user.getUsername().equals("张三")&&user.getPassword().equals("123")){
            return "baidu";
        }else{
```

```
            return "mail163";
        }
    }
```

（2）修改 springmvc.xml 配置文件，增加下面的配置，注意还要设置视图解释器的优先级，因为新添加了一个 BeanNameViewResolver 视图解释器，配置文件中因而一共存在两种视图解释器，会有冲突，配置了优先级则解决了冲突问题。首先匹配优先级高的视图解释器，若找不到再匹配优先级低的视图解释器。具体配置优先级的方法见下面代码：

```xml
<!-- 配置内部视图解析器 -->
<bean class="org.springframework.web.servlet.view.InternalResourceViewResolver">
    <!--逻辑视图前缀-->
    <property name="prefix" value="/WEB-INF/jsp/"></property>
    <!--逻辑视图后缀，匹配模式：前缀+逻辑视图+后缀，形成完整路径名-->
    <property name="suffix" value=".jsp"></property>
</bean>
<!-- 配置 BeanNameViewResolver 类型的视图解释器 -->
<bean class="org.springframework.web.servlet.view.BeanNameViewResolver">
    <property name="order" value="5"/>
</bean>
<bean id="baidu" class="org.springframework.web.servlet.view.RedirectView">
    <property name="url" value="http://www.baidu.com"/>
</bean>
<bean id="mail163" class="org.springframework.web.servlet.view.RedirectView">
    <property name="url" value="http://email.163.com"/>
</bean>
```

> 设置优先级为 5，比内部视图解释器要高，优先使用

（3）运行测试，登录成功后跳转到了百度，失败后跳转到了 163 邮箱。

10.7.2 使用 Model 参数

使用 ModelAndView 对象作为控制器方法的返回值类型，可以很好地同时解决跳转路径和携带的数据问题，但 String 作为控制器方法的返回值类型只解决了跳转路径问题，如果跳转到目标页面同时还要携带数据怎么办呢？有 3 个解决方案：①使用方法中的 Model 参数；②使用方法中的 HttpRequestServlet 对象；③使用方法中的 HttpSession 对象。

在参数中使用 Model，可以把数据封装到 Model 对象中，进而再跳转到目标页面，目标页面使用 EL 表达式读取出来。上面提到，Model 与 HttpServletRequest、HttpSession 三者均可以携带数据从后台到前台。

项目案例：登录页面，登录成功后跳转到登录成功页面，并把后台传来的用户名显示到此页面。（项目源码参见配套资源：第 10 章/返回 String 并携带数据/springmvc1201）

实现步骤：

（1）复制项目 springmvc12 为 springmvc1201，修改 UserController 中的 doLogin 方法，添加 Model 参数，代码如下：

```java
//在参数中添加model,实现数据传回客户端
@RequestMapping("/dologin.do")
public String doLogin(User user,Model model){
    if(user.getUsername().equals("张三")&&user.getPassword().equals("123")){
        model.addAttribute("user",user.getUsername());
```

```
            return "success";
    }else{
            return "login";
    }
}
```

（2）修改 WEB-INF/jsp 下的 success.jsp 如下：
```
<body>
<h1>登录成功,欢迎${username }光临</h1>
</body>
```

（3）运行测试，登录成功的情况下，数据可以传递到前台。

10.7.3　使用 HttpSerlvetRequest 参数

可以用 HttpSerlvetRequest 作为参数，实现数据从后台传递到前台。只须将上述代码修改如下。（项目源码参见配套资源：第 10 章/返回 String 并携带数据/springmvc1201）

```
//在参数中添加HttpSerlvetRequest,实现数据传回客户端
    @RequestMapping("/dologin.do")
    public String doLogin(User user,HttpServletRequest request){
        if(user.getUsername().equals("张三")&&user.getPassword().equals("123")){
            request.setAttribute("user",user.getUsername());
            return "success";
        }else{
            return "login";
        }
    }
```

结果同上。

10.7.4　使用 HttpSession 参数

可以用 HttpSession 作为参数，实现数据从后台传递到前台。只须将上述代码修改如下。（项目源码参见配套资源：第 10 章/返回 String 并携带数据/springmvc1201）

```
//在参数中添加HttpSession,实现数据传回客户端
    @RequestMapping("/dologin.do")
    public String doLogin(User user,HttpSession session){
        if(user.getUsername().equals("张三")&&user.getPassword().equals("123")){
            session.setAttribute("user",user.getUsername());
            return "success";
        }else{
            return "login";
        }
    }
```

10.8　控制器方法返回 void 类型

void 表示没有返回值，没有返回值要解决两个问题：一是如何跳转，二是数据如何传回前台。

10.8.1 使用 ServletAPI 参数

可以在处理器方法的参数中放入 ServletAPI 参数,来完成数据的传递及资源跳转。具体来说,可在方法参数中放入 HttpServletRequest 或 HttpSession,使方法可以直接将数据放入 request、session 的域中,也可通过 request.getServletContext()获取到 ServletContext,从而将数据放入 application 的域中。

方法参数中放入 HttpServletRequest 或 HttpServletResponse,可以完成请求转发与重定向,但需要注意,重定向是无法访问/WEB-INF/下资源的。

请求转发代码:
`request.getRequestDispatcher("目标页面").forward(request,response);`

重定向代码:
`response.sendRedirect("目标页面");`

项目案例:将项目 springmvc12 复制为 springmvc13,登录方法改为无返回值,实现一样能正常登录。(项目源码参见配套资源:第 10 章/ServletAPI 接收数据并跳转/springmvc13)

关键步骤:

(1)修改项目 springmvc13 的 doLogin()方法如下,类似学习 JSP+Servlet 时的技术:

```
//无返回值的情形,由servlet来实现跳转,HttpServletRequest携带数据回客户端
@RequestMapping("/dologin.do")
public void doLogin(HttpServletRequest request,
        HttpServletResponse response,User user) throws IOException, ServletException{
    if(user.getUsername().equals("张三")&&user.getPassword().equals("123")){
        request.setAttribute("user",user);
request.getRequestDispatcher("/WEB-INF/jsp/success.jsp").forward(request,response);;
    }else{
        response.sendRedirect("login.do");
    }
}
```

(2)其他不变,测试登录成功与失败的情形。

10.8.2 Ajax 响应

若处理器对请求处理后,无须跳转到其他任何资源,则可以让处理器方法返回 void。例如,对于 Ajax 的异步请求的响应,其请求完成后往往还停留在原来页面,无须跳转,可以让处理器的方法返回 void。难点是参数如何从客户端向服务端传递,以及处理器处理完成后的数据如何返回客户端。通过下面的项目案例来了解整个过程。

项目案例:在项目中添加一个页面,利用 Ajax 实现单击按钮弹出服务端回传来的用户名与密码。(项目源码参见配套资源:第 10 章/ajax 响应/springmvc14)

实现步骤:

(1)将项目 springmvc13 复制为 springmvc14,导入下述 JAR 包。

本项目中服务端向浏览器传回的是 JSON(JavaScript Object Notation,JS 对象符号)格式数据,需要使用一个工具类将字符串包装为 JSON 格式,故需导入 JSON 的 JAR 包。相关的 JAR 共 5 个:
commons-beanutils-1.8.0.jar、commons-collections-3.2.1.jar、commons-lang-2.4.jar、ezmorph-1.0.6.jar、

json-lib-2.3-jdk15.jar。

（2）引入 jQuery 库。

由于本项目要使用 jQuery 的 ajax()方法提交 Ajax 请求，所以项目中需要引入 jQuery 的库。在 WebContent 下新建一个 Folder（文件夹），将其命名为 js，并将 jquery-1.11.2.min.js 文件放入其中。

（3）在 WebContent 下创建 index.jsp 页面，添加按钮，处理 Ajax 请求，代码如下：

```jsp
<%@ page language="java" contentType="text/html; charset=utf-8"
    pageEncoding="utf-8"%>
<!DOCTYPEhtmlPUBLIC"-//W3C//DTDHTML4.01Transitional//EN" "http://www.w3.org/TR/html4/loose.dtd">
<html>
<head>
<meta http-equiv="Content-Type" content="text/html; charset=ISO-8859-1">
<title>Insert title here</title>
<script type="text/javascript" src="js/jquery-1.11.2.min.js"></script>
<script type="text/javascript">
$(document).ready(function(){
    $("#btn1").click(function(){
        $.ajax({
            url:"user/doajax.do",
            data:{
                username:"张三",
                password:"123"
            },
            success:function(data){
                var jsonobject=JSON.parse(data);
                alert("用户名:"+jsonobject.username+"密码:"+jsonobject.password);
            }
        });
    });
});
</script>
</head>
<body>
<input id="btn1" type="button" value="点我"/>
</body>
</html>
```

从 Ajax 方法可以看出，客户端发送给服务端的数据格式是 JSON。接下来是服务端如何接收 JSON 数据的问题。

（4）在控制器 UserController 中添加 doAjax()方法。

处理器对于 Ajax 请求中所提交的 JSON 参数数据，可以使用逐个接收的方式，也可以以对象的方式整体接收，只要保证 Ajax 请求参数与接收的对象类型属性同名，则跟其他方式提交的参数处理并无差别。这里以对象方式整体接收为例，代码如下：

```java
@RequestMapping("/doajax.do")
    public void doAjax(HttpServletRequest request,
            HttpServletResponse response,User user) throws IOException{
        Map<String,String> map=new HashMap<String,String>();
        map.put("username", user.getUsername());
        map.put("password", user.getPassword());
        //把map对象转换为JSONObject对象
```

```
            JSONObject jsonObject=JSONObject.fromObject(map);
            //把 JSONObject 对象转换为 JSON 格式的字符串
            String jsonStr=jsonObject.toString();
            response.setCharacterEncoding("utf-8");
            PrintWriter out=response.getWriter();
            out.print(jsonStr);
            out.close();
        }
```
处理器方法处理完客户端传来的数据后,以 PrintWriter 对象的 print 方法直接输出 json 格式的字符串,返回给前端调用者(Ajax 方法)。

(5)运行测试,结果如图 10.17 所示。

图 10.17　Ajax 响应结果页

10.9　控制器方法返回 Object 类型

处理器方法也可以返回 Object 对象,返回的这个 Object 对象不再作为逻辑视图出现,而是作为直接在页面显示的数据出现。

控制器方法欲返回 Object 对象,需要使用@ResponseBody 注解,将转换后的 JSON 数据放入响应体中。返回 Object 数据,一般都是将数据转化成为 JSON 对象后传递给浏览器页面的,这个由 Object 转换为 JSON 的工作,由 Jackson 工具完成,故需要导入 Jackson 的相关 JAR 包(具体见下面案例)。

根据 Object 对象实际的类型,又可分别返回下述类型:数值型对象、字符串型对象、Bean 对象、Map 集合、List 集合。

项目案例:在 index.jsp 页面中创建几个按钮,单击后分别返回数值型对象、字符串型对象、Bean 对象、Map 集合、List 集合。(项目源码参见配套资源:第 10 章/返回 Object/springmvc15)

实现步骤:

(1)复制项目 springmvc14 为 springmvc15,导入下列 JAR 包:jackson-annotations-2.4.0.jar、jackson-core-2.4.1.jar、jackson-databind-2.4.1.jar。

(2)在 springmvc.xml 中注册注解驱动:
```
<!-- 注册 springmvc 的注解驱动 -->
<mvc:annotation-driven/>
```
添加注解驱动的作用是使 JSON 字符串能自动转换为实体类。

(3)修改 index.jsp,添加若干个按钮和对应的 Ajax 方法,代码如下:
```
<%@ page language="java" contentType="text/html; charset=utf-8"
    pageEncoding="utf-8"%>
<!DOCTYPEhtmlPUBLIC"-//W3C//DTDHTML4.01Transitional//EN" "http://www.w3.org/TR/html4/
```

```
   loose.dtd">
   <html>
   <head>
   <meta http-equiv="Content-Type" content="text/html; charset= utf-8">
   <title>Insert title here</title>
   <script type="text/javascript" src="js/jquery-1.11.2.min.js"/>
   <script type="text/javascript">
   $(function(){
      $("#btn1").click(function(){
       $.ajax({
           url:"user/doajax.do",
           data:{username:"张三",password:"123"},
           success:function(data){
               var jsonobject=JSON.parse(data);
               alert("用户名:"+jsonobject.username+"密码:"+jsonobject.password);
           }
       });
      });
      $("#btn2").click(function(){
       $.ajax({
           url:" user/doajax2.do",
           success:function(data){
               alert(data);
           }
       });
      });
      $("#btn3").click(function(){
       $.ajax({
           url:" user/doajax3.do",
           success:function(data){
               alert(data);
           }
       });
      });
      $("#btn4").click(function(){
       $.ajax({
           url:" user/doajax4.do",
           success:function(data){
               alert("用户名:"+data.username+"密码:"+data.password);
           }
       });
      });
      $("#btn5").click(function(){
       $.ajax({
           url:" user/doajax5.do",
           success:function(data){
               alert("用户名:"+data.user1.username+"密码:"+data.user1.password);
               alert("用户名:"+data.user2.username+"密码:"+data.user2.password);
           }
       });
      });
      $("#btn6").click(function(){
       $.ajax({
           url:" user/doajax6.do",
```

```
            success:function(data){
                $(data).each(function(index){
                    alert("用户名:"+data[index].username+"密码:"+data[index].password)
                });
            }
        });
    });
});
</script>
</head>
<body>
<input id="btn1" type="button" value="点我"/>
<input id="btn2" type="button" value="返回数值"/>
<input id="btn3" type="button" value="返回字符"/>
<input id="btn4" type="button" value="返回对象"/>
<input id="btn5" type="button" value="返回 Map"/>
<input id="btn6" type="button" value="返回 List"/>
</body>
</html>
```

（4）在 User controller 中添加方法 doAjax2()实现返回数值型对象，代码如下：
```
@RequestMapping("/doajax2.do")
@ResponseBody
public Object doAjax2(){
    return 520.1314;
}
```
这里也可直接返回 Object 实际代表的 double 类型，效果一样，代码如下：
```
@RequestMapping("/doajax2.do")
@ResponseBody
public double doAjax2(){
    return 520.1314;
}
```
@ResponseBody 指将 JSON 字符串作为响应处理。

（5）在 User controller 中添加方法 doAjax3()实现返回字符串型对象，代码如下：
```
@RequestMapping("/doajax3.do")
@ResponseBody
public Object doAjax3(){
    return "Hello SpringMVC";
}
```
这里也可直接返回 Object 实际代表的 String 类型，效果一样，不过很显然必须有@ResponseBody 注解，否则返回的字符串会被解释为视图，代码如下：
```
@RequestMapping("/doajax3.do")
@ResponseBody
public String doAjax3(){
    return "Hello SpringMVC";
}
```

（6）在 User controller 中添加方法 doAjax4()实现返回 Bean 类型对象，代码如下：
```
@RequestMapping("/doajax4.do")
@ResponseBody
public Object doAjax4(){
    User user=new User("张三","123");
```

```
        return user;
    }
```

这里也可直接返回 Object 实际代表的 User 类型，效果一样，一般直接用 Bean 类型，代码如下：

```
@RequestMapping("/doajax4.do")
@ResponseBody
public User doAjax4(){
    User user=new User("张三","123");
    return user;
}
```

（7）在 User controller 中添加方法 doAjax5()实现返回 Map 类型对象，代码如下：

```
@RequestMapping("/doajax5.do")
@ResponseBody
public Object doAjax5(){
    User user1=new User("张三","111");
    User user2=new User("李四","222");
    Map<String,User> map=new HashMap<String,User>();
    map.put("user1", user1);
    map.put("user2", user2);
    return map;
}
```

（8）在 User controller 中添加方法 doAjax6()实现返回 List 类型对象，代码如下：

```
@RequestMapping("/doajax6.do")
@ResponseBody
public Object doAjax6(){
    User user1=new User("张三","111");
    User user2=new User("李四","222");
    User user3=new User("王五","333");
    List<User> list=new ArrayList<User>();
    list.add(user1);
    list.add(user2);
    list.add(user3);
    return list;
}
```

（9）运行测试，效果如图 10.18～图 10.20 所示。

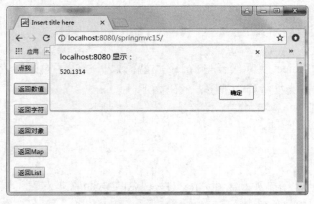

图 10.18　返回数值对象

图 10.19　返回字符串

图 10.20　返回 Bean 对象

其他 Map、List 类似。

10.10　Ajax/JSON 专项突破

Ajax 请求及响应同时采用 JSON 格式的数据进行参数的传递是 Web 开发中最常见的方式之一，下面深入剖析其各种应用情形。

10.10.1　服务端接收对象返回 JSON 字符串

项目案例：添加按钮，单击按钮，用 Ajax 发送数据到后台，并从后台获取数据。（项目源码参见配套资源：第 10 章/ajax 与 json/springmvc16）

关键步骤：

（1）复制项目 springmvc15 为 springmvc16，在 index.jsp 页面中新添加一个按钮及对应的 Ajax 方法。添加的按钮如下：

```
<input id="btn7" type="button" value="Ajax 发送 JSON,服务端接收 Bean 返回 JSON 字符串"/>
<br/><br/>
```

添加的 Ajax 方法如下：

```
$("#btn7").click(function(){
    $.ajax({
```

```
                url:"user/doajax7.do",
                data:{
                    username:"张三",
                    password:"123"
                },
                success:function(data){
                    var jsonObject=JSON.parse(data); //用于将服务端传回来的JSON字符串解释转化
```
为JSON对象
```
                    alert("用户名:"+jsonObject.username+"密码:"+jsonObject.password);
                }
            });
        });
```
（2）在控制器中添加方法doAjax7：
```
@RequestMapping("/doajax7.do")
public void doAjax7(HttpServletRequest request,
        HttpServletResponse response,User user) throws ServletException,IOException{
    Map<String,String> map=new HashMap<String,String>();
    map.put("username", user.getUsername());
    map.put("password", user.getPassword());
    //把map对象转换为JSONObject对象
    JSONObject jsonObject=JSONObject.fromObject(map);
    //把JSONObject对象转换为JSON格式的字符串
    String jsonStr=jsonObject.toString();
    response.setCharacterEncoding("utf-8");
    PrintWriter out=response.getWriter();
    out.print(jsonStr);
    out.close();
}
```
该方法中的user参数用于接收客户端Ajax方法传来的JSON格式数据。

（3）测试运行，结果如图10.21所示。

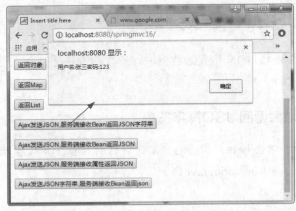

图 10.21 JSON 运行结果

10.10.2 服务端接收 Bean 返回 JSON 对象

项目案例：添加按钮，单击按钮，用 Ajax 发送数据到后台，并从后台获取数据。（项目源码参见配套资源：第 10 章/ajax 与 json/springmvc16）

关键步骤：
（1）在 index.jsp 页面中新添加一个按钮及对应的 Ajax 方法。添加的按钮如下。
```
<input id="btn8" type="button" value="Ajax 发送 JSON,服务端接收 Bean 返回 JSON"/><br/><br/>
```
添加的 Ajax 方法如下：
```
$("#btn8").click(function(){
        $.ajax({
            url:"user/doajax8.do",
            data:{
                username:"张三",
                password:"123"
            },
            success:function(data){
                alert("用户名:"+data.username+"密码:"+data.password);
            }
        });
    });
```
（2）在控制器中添加方法 doAjax8 如下：
```
@RequestMapping("/doajax8.do")
@ResponseBody
public User doAjax8(User user) throws ServletException,IOException{
    return user;
}
```
该方法中的 user 参数用于接收客户端 Ajax 方法传来的 JSON 对象格式数据。返回的 user 对象会被封装为 JSON 格式数据，因为有@ResponseBody 注解。

（3）运行测试，结果同前，如图 10.22 所示。

图 10.22　JSON 运行结果

10.10.3　服务端接收属性返回 JSON 对象

项目案例：添加按钮，单击按钮，用 Ajax 发送数据到后台，并从后台获取数据。（项目源码参见配套资源：第 10 章/ajax 与 json/springmvc16）

关键步骤：
（1）在 index.jsp 页面中新添加一个按钮及对应的 Ajax 方法。
添加的按钮如下：

```
<input id="btn9" type="button" value="Ajax发送JSON,服务端接收属性返回JSON"/><br/><br/>
```
添加的 Ajax 方法如下：
```
$("#btn9").click(function(){
    $.ajax({
        url:"user/doajax9.do",
        data:{
            username:"张三",
            password:"123"
        },
        success:function(data){
            alert("用户名:"+data.username+"密码:"+data.password);
        }
    });
});
```
（2）在控制器中添加方法 doAjax9 如下：
```
@RequestMapping("/doajax9.do")
@ResponseBody
publicUserdoAjax9(Stringusername,Stringpassword)throws ServletException,IOException{
    User user=new User(username, password);
    return user;
}
```
（3）运行测试，结果同前，如图 10.23 所示。

图 10.23　JSON 运行结果

10.10.4　客户端发送 JSON 字符串返回 JSON 对象

项目案例：添加按钮，单击按钮，用 Ajax 发送数据到后台，并从后台获取数据。（项目源码参见配套资源：第 10 章/ajax 与 json/springmvc16）

关键步骤：

（1）在 index.jsp 页面中新添加一个按钮及对应的 Ajax 方法。

添加的按钮如下：
```
<input id="btn10" type="button" value="Ajax 发送 JSON 字符串,服务端接收 Bean 返回 json"/><br/><br/>
```
添加的 Ajax 方法如下：

```
$("#btn10").click(function(){
    var jsonObj={username:"张三",  password:"123"   };//定义一个 JSON 对象
    var jsonStr=JSON.stringify(jsonObj);//将 JSON 对象转化为字符串
        $.ajax({
            url:"user/doajax10.do",
            data: jsonStr,
            type:"post",
            contentType:"application/json;charset=UTF-8",
            success:function(data){
                alert("用户名:"+data.username+"密码:"+data.password);
            }
        });
    });
```

【注意】要用 post 方式发送字符串，contentType:"application/json;charset=UTF-8"表示告诉服务器发送请求的数据格式为 JSON。

（2）在控制器中添加方法 doAjax10 如下：
```
@RequestMapping("/doajax10.do")
@ResponseBody
public User doAjax10(@RequestBody User user){
    return user;
}
```
这表示将前端请求体中的 JSON 格式数据绑定到对应的形式参数上。

（3）运行测试，结果同前，如图 10.24 所示。

图 10.24　JSON 运行结果

10.10.5　数据接收与返回的格式限制

控制器方法通过 consumes="application/json"限制前台传递过来的数据格式必须是 JSON，通过属性 produces="application/json"设置返回的数据要转成 JSON 对象并回传给客户端，当然，也可根据需要设置为其他数据格式。

项目案例：添加按钮，单击按钮，用 Ajax 发送数据到后台，并从后台获取数据。（项目源码参见配套资源：第 10 章/ajax 与 json/springmvc16）

关键步骤：

（1）在 index.jsp 中添加按钮与 Ajax 方法如下：

```
$("#btn11").click(function(){
    var myJson={"username":"张三","password":"123"};
    var jsonStr=JSON.stringify(myJson);
    $.ajax({
        url:"user/doajax11.do",
        data:jsonStr,
        contentType:"application/json",
        dataType:"json",
        type:"post",
        success:function(data){
            alert("用户名:"+data.username+"密码:"+data.password);
        }
    });
});
```

dataType:"json"表示告诉服务器需要接收类型为 json 的响应数据，也可以省略，页面会自动识别响应数据的类型，即自动判断是返回 XML、JSON、script，还是返回 String。

```
<input id="btn11" type="button" value="Ajax 发送 JSON 字符串,服务端接收 JSON 对象"/>
```

（2）UserController 添加 doAjax11()方法如下：

```
@RequestMapping(value="/doajax11.do",consumes="application/json",produces="application/json")
    @ResponseBody
    public User doAjax11(@RequestBody User user){
        System.out.println(user.getUsername());//控制台测试数据,无实际作用,可不要
        return user;
    }
```

（3）运行测试，结果如图 10.25 所示。

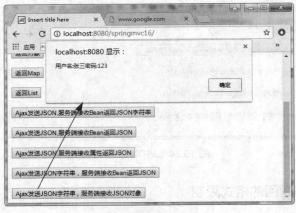

图 10.25　JSON 运行结果

10.10.6　直接输出响应字符串

项目案例：添加按钮，单击按钮，用 Ajax 发送数据到后台，并从后台获取数据。（项目源码参见配套资源：第 10 章/ajax 与 json/springmvc16）

第 10 章 Spring MVC 注解式开发

关键步骤：

（1）在 index.jsp 页面中添加按钮及 Ajax 方法。添加 Ajax 方法代码如下：

```
$("#btn12").click(function(){
    $.ajax({
        url:"user/doajax12.do",
        data:"name=Hello MVC!!!",
        success:function(data){
            alert(data);
        }
    });
});
```

添加按钮如下：

```
<input id="btn12" type="button" value="直接输出响应字符串 "/>
```

（2）在 UserController 方法中添加方法 doAjax12()，代码如下：

```
//直接输出响应字符串
    @RequestMapping("/doajax12.do")
    public void doAjax12(String name,HttpServletResponse response) throws IOException{
        response.setContentType("text/html");
        response.setCharacterEncoding("utf-8");
        PrintWriter out=response.getWriter();
        out.print(name);
        out.flush();
        out.close();
    }
```

类似地可以从数据库查找出数据响应给前台。

（3）运行测试，效果如图 10.26 所示。

图 10.26 JSON 运行结果

通过上述多种形式的交互，可以总结出客户端发送 JSON 对象格式最方便，发送 JSON 字符串比较麻烦。服务端返回 JSON 对象，客户端处理起来也更方便。

上机练习

题目：用户注册后，用 Ajax/JSON 方式实现数据的显示。

思考题

1. 参数从客户端传到服务端有哪些方式?
2. 参数从服务端传到客户端有哪些方式?
3. 如果方法的返回值为 String，有哪些办法可实现向前台传递数据?
4. 总结 Ajax 与 JSON 的数据传递过程。

第 11 章 Spring MVC 关键技术

本章目标

- ✧ 掌握转发和重定向原理与操作的方法
- ✧ 掌握异常处理的方法
- ✧ 会进行数据验证
- ✧ 会进行类型转换
- ✧ 掌握文件上传的方法
- ✧ 掌握文件下载的方法
- ✧ 会使用拦截器
- ✧ 会配置静态资源访问

11.1 转发与重定向

当处理器完成请求处理后向其他资源跳转时，有两种跳转方式：请求转发与重定向。根据跳转的资源类型，可将跳转分为两类：跳转到页面与跳转到其他处理器。请求转发的页面可以是 WEB-INF 中的页面，但重定向的页面不能是 WEB-INF 中的页面，因为重定向相当于用户重新发出一次请求，而用户是不可以直接访问 WEB-INF 中的资源的。

11.1.1 请求转发到其他页面

当处理器方法返回 ModelAndView 时，跳转到指定的 ViewName，默认情况下使用的是请求转发，当然也可显式地进行请求转发。此时，需要在 setViewName()指定的视图前添加 forward 关键字，一旦添加了 forward 关键字，控制器方法返回的视图名称就不会再与视图解析器中的前缀及后缀进行拼接，所以必须写出相对项目根的完整路径才能返回正确的视图。

当通过请求转发跳转到目标资源（页面或 Controller）时，若需要目标资源传递数据，可以使用 HttpRequestServlet 及 HttpSession，还可以将数据存放在 ModelAndView 中的 Model 中。目标页面则通过 EL 表达式来访问该数据。下面的案例演示了使用 ModelAndView 的情形。

项目案例：用户注册完毕后，显示用户的注册信息。（项目源码参见配套资源：第 11 章/转发与重定向/springmvc17）

关键步骤：

（1）新建 Dynamic Web Project 项目 springmvc17，添加 JAR 包，按之前学习的步骤搭建框架，web.xml 和 springmvc.xml 的配置参考之前的案例（可参考 springmvc8）。在 WEB-INF/jsp 下新建 register.jsp 和 info.jsp 页面。

register.jsp 代码如下：

```
<%@ page language="java" contentType="text/html; charset=utf-8"
    pageEncoding="utf-8"%>
<!DOCTYPE html PUBLIC "-//W3C//DTD HTML 4.01 Transitional//EN" "http://www.w3.org/TR/html4/loose.dtd">
<html>
<head>
<meta http-equiv="Content-Type" content="text/html; charset=ISO-8859-1">
<title>Insert title here</title>
</head>
<body>
用户注册
<form action="doregister.do">
姓名:<input type="text" name="username"/><br/>
密码:<input type="text" name="password"/><br/>
<input type="submit" value="注册"/>
</form>
</body>
</html>
```

info.jsp 代码如下：

```
<%@ page language="java" contentType="text/html; charset=utf-8"
    pageEncoding="utf-8"%>
```

```
<!DOCTYPE html PUBLIC "-//W3C//DTD HTML 4.01 Transitional//EN" "http://www.w3.org/TR/html4/loose.dtd">
<html>
<head>
<meta http-equiv="Content-Type" content="text/html; charset=ISO-8859-1">
<title>Insert title here</title>
</head>
<body>
用户注册信息<br/>
用户名:${user.username}<br/>
密码:${user.password}<br/>
</body>
</html>
```

（2）新建包 com.lifeng.entity，新建实体类 User，代码如下：

```
package com.lifeng.entity;
public class User {
    String username;
    String password;
} //省略 getter,setter 方法
```

（3）新建包 com.lifeng.controller，在包下新建 UserController 控制器，方法如下：

```
@RequestMapping("/register.do")
public String register(){
    return "register";
}
@RequestMapping("/doregister.do")
public ModelAndView doRegister(User user){
    ModelAndView mv=new ModelAndView();
    mv.addObject("user",user);
    mv.setViewName("forward:/WEB-INF/jsp/info.jsp");
    return mv;
}
```

转发的目标页面，需要写全路径

（4）运行测试，输入 http://localhost:8080/springmvc17/user/register.do，注册和转发页面如图 11.1 和图 11.2 所示。

图 11.1　注册页面

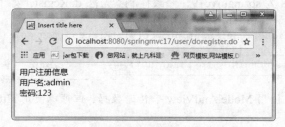

图 11.2　转发页面

11.1.2 请求转发到其他控制器

当前控制器的处理方法在处理完毕后也可以不返回视图,而是转发给下一个控制器方法继续进行处理。

项目案例:用户注册成功后,转发给其他方法,由其他方法返回视图显示当前用户的基本信息。(项目源码参见配套资源:第 11 章/转发与重定向/springmvc18)

关键步骤:

(1)复制 springmvc17 为 springmvc18,修改或添加 UserController 中的两个方法:

```
@RequestMapping("/doregister.do")
public ModelAndView doRegister(User user){
    ModelAndView mv=new ModelAndView();
     mv.addObject("user",user);
    mv.setViewName("forward:second.do");
    return mv;
}
@RequestMapping("/second.do")
public ModelAndView doSecond(User user){
    ModelAndView mv=new ModelAndView();
    mv.addObject("user",user);
    mv.setViewName("forward:/WEB-INF/jsp/info.jsp");
    return mv;
}
```

参数传递方向

可以发现,参数仍然可以在两个方法之间传递,第一个方法把参数存进 ModelAndView,第二个方法用同名形式参数接收。

mv.setViewName("forward:second.do");这行代码实现转发到另一个方法 second.do 继续处理。

(2)运行测试。输入 http://localhost:8080/springmvc18/user/register.do,结果同前。

11.1.3 返回 String 时的请求转发

当处理器方法返回 String 时,该 String 即为要跳转的视图。必须在其前面加上前缀 forward:,显式地指定跳转方式为请求转发。视图解析器将不会对其进行前缀及后缀的拼接,该 String 中的路径需是完整路径。

请求转发的目标资源无论是一个页面,还是一个 Controller,其用法都一样。

项目案例:用户注册成功后,转发给其他方法,由其他方法返回视图显示当前用户的基本信息。(项目源码参见配套资源:第 11 章/转发与重定向/springmvc19)

关键步骤:

(1)复制 springmvc18 为 springmvc19,修改控制器方法 doRegister 如下:

```
@RequestMapping("/doregister.do")
public String doRegister(User user,HttpServletRequest request){
     request.setAttribute("user", user);
     return "forward:/WEB-INF/jsp/info.jsp";
}
```

完整路径

【注意】这种情况不能使用 ModelAndView 来传递数据,但可以使用 HttpServletRequest 等来传递数据。

（2）运行测试。输入 http://localhost:8080/springmvc19/user/register.do，结果同前。

11.1.4 请求重定向到其他页面

在重定向时，请求参数不能通过 HttpServletRequest 向目标资源中传递。可以通过以下方式之一来传递请求参数。

1. 通过 ModelAndView 中的 Model 携带参数

当 ModelAndView 中的 Model 存入数据后，视图解析器 InternalResourceViewResolver 会将 map 中的 key 与 value 以请求参数的形式放到请求的 URL 后。注意事项如下。

（1）放入 Model 中的 value 只能是基本数据类型与 String，不能是自定义类型的对象数据。原因是视图解析器会将 Map 的 value 放入 URL 后作为请求参数传递出去，任何类型的 value 都会变为 String。

（2）在重定向的页面中是无法从 request 中读取数据的。但由于 map 中的 key 与 value，以请求参数的形式放到了请求的 URL 后，所以页面可以通过 EL 表达式中的请求参数 param 读取。

（3）重定向的页面不能是/WEB-INF 下的页面。因为重定向相当于客户端发出一次新的请求，而客户端是不可以请求/WEB-INF 下的资源的。

项目案例：用户登录成功后，通过重定向页面实现登录后显示用户信息。（项目源码参见配套资源：第 11 章/转发与重定向/springmvc20）

关键步骤：

（1）复制项目 springmvc19 为 springmvc20，在 WebContent 下创建页面 show.jsp，复制之前的 login.jsp 页面。

Show.jsp 代码如下：

```
<%@ page language="java" contentType="text/html; charset=utf-8"
    pageEncoding="utf-8"%>
<!DOCTYPE html PUBLIC "-//W3C//DTD HTML 4.01 Transitional//EN" "http://www.w3.org/TR/html4/loose.dtd">
<html>
<head>
<meta http-equiv="Content-Type" content="text/html; charset= utf-8">
<title>Insert title here</title>
</head>
<body>
登录用户信息<br/>
用户名:${param.username}<br/>
密码:${param.password}<br/>
</body>
</html>
```

【注意】 这里用到了 param 对象。

Login.jsp 代码如下：

```
<%@ page language="java" contentType="text/html; charset=utf-8"
    pageEncoding="utf-8"%>
<!DOCTYPE html PUBLIC "-//W3C//DTD HTML 4.01 Transitional//EN" "http://www.w3.org/TR/html4/loose.dtd">
<html>
```

```
<head>
<meta http-equiv="Content-Type" content="text/html; charset=utf-8">
<title>Insert title here</title>
</head>
<body>
用户登录
<form action="dologin.do">
姓名:<input type="text" name="username"/><br/>
密码:<input type="text" name="password"/><br/><br/>
<input type="submit" value="登录"/>
</form>
</body>
</html>
```

（2）在 UserController 中添加 doLogin 方法，代码如下：

```
@RequestMapping("/login.do")
public String login(){
    return "login";
}
@RequestMapping("/dologin.do")
public ModelAndView doLogin(User user){
    ModelAndView mv=new ModelAndView();
    mv.addObject("username",user.getUsername());
    mv.addObject("password",user.getPassword());
    mv.setViewName("redirect:/show.jsp");
    return mv;
}
```

重定向到 show.jsp 页面

（3）测试运行，输入 http://localhost:8080/springmvc20/user/login.do。

再次测试：如果在 show.jsp 页面删除 param，能否接收到数据。

2. 使用 HttpSession 携带参数

项目案例：用户登录成功后，通过重定向页面实现登录后显示用户信息。（项目源码参见配套资源：第 11 章/转发与重定向/springmvc21）

关键步骤：

（1）在 WebContent 下创建 show2.jsp，代码如下：

```
<%@ page language="java" contentType="text/html; charset=utf-8"
    pageEncoding="utf-8"%>
<!DOCTYPE html PUBLIC "-//W3C//DTD HTML 4.01 Transitional//EN" "http://www.w3.org/TR/html4/loose.dtd">
<html>
<head>
<meta http-equiv="Content-Type" content="text/html; charset= utf-8">
<title>Insert title here</title>
</head>
<body>
登录用户信息<br/>
用户名:${user.username}<br/>
密码:${user.password}<br/>
</body>
</html>
```

（2）修改 doLogin 方法，代码如下：

```
@RequestMapping("/dologin.do")
public ModelAndView doLogin(User user,HttpSession session){
    ModelAndView mv=new ModelAndView();
    session.setAttribute("user", user);
    mv.setViewName("redirect:/show2.jsp");
    return mv;
}
```

（3）测试运行，输入 http://localhost:8080/springmvc21/user/login.do。

11.1.5 请求重定向到其他控制器

重定向到其他 Controller 方法时，携带参数可以采用前面的其中一个方式。而目标 Controller 接收这些参数，也有多种方式。

1. 通过 ModelAndView 的 Model 携带参数

目标 Controller 在接收这些参数时，只要保证目标 Controller 的方法形参名称与发送 Controller 发送的参数名称相同即可接收。当然，目标 Controller 也可以进行参数的整体接收。只要保证参数名称与目标 Controller 接收参数类型的类的属性名相同即可。

项目案例：用户登录成功后，通过重定向页面实现登录后显示用户信息。（项目源码参见配套资源：第 11 章/转发与重定向/springmvc22）

关键步骤：

（1）修改 doLogin 方法，添加两个目标方法，代码如下：

```
@RequestMapping("/dologin.do")
public ModelAndView doLogin(User user){
    ModelAndView mv=new ModelAndView();
    mv.addObject("username",user.getUsername());
    mv.addObject("password",user.getPassword());
    //第1次测试
    mv.setViewName("redirect:second.do");
    //第2次测试
    //mv.setViewName("redirect:third.do");
    return mv;
}
//整体接收
@RequestMapping("/second.do")
public ModelAndView doSecond(User user){
    ModelAndView mv=new ModelAndView();
    mv.addObject("username",user.getUsername());
    mv.addObject("password",user.getPassword());
    mv.setViewName("redirect:/show.jsp");
    return mv;
}
//逐个参数接收
@RequestMapping("/third.do")
public ModelAndView doThird(String username,String password){
    ModelAndView mv=new ModelAndView();
    mv.addObject("username",username);
    mv.addObject("password",password);
    mv.setViewName("redirect:/show.jsp");
    return mv;
}
```

（2）测试运行，输入 http://localhost:8080/springmvc22/user/login.do。

注释掉 mv.setViewName("redirect:second.do")，添加 mv.setViewName("redirect:third.do")再次测试。观察两次结果是否相同。

2. 使用 HttpSession 携带参数

项目案例：用户登录成功后，通过重定向页面实现登录后显示用户信息。（项目源码参见配套资源：第 11 章/转发与重定向/springmvc23）

关键步骤：

（1）复制项目 springmvc22 为 springmvc23，修改处理器类 UserController，代码如下：

```
@RequestMapping("/dologin.do")
public ModelAndView doLogin(User user,HttpSession session){
    session.setAttribute("user", user);
    ModelAndView mv=new ModelAndView();
    mv.setViewName("redirect:fourth.do");
    return mv;
}
@RequestMapping("/fourth.do")
public ModelAndView doFifth(HttpSession session){
    User user=(User) session.getAttribute("user");
    ModelAndView mv=new ModelAndView();
    mv.addObject("username",user.getUsername());
    mv.addObject("password",user.getPassword());
    mv.setViewName("redirect:/show.jsp");
    return mv;
}
```

（2）测试运行，输入 http://localhost:8080/springmvc23/user/login.do。

11.1.6 返回 String 时的重定向

可以重定向到页面，也可以重定向到其他控制器方法。当处理器的方法返回类型为 String 时，在字符串中添加前缀 redired:即可实现重定向。如果还要传递参数，可以通过 URL、HttpSession、Model 携带参数等多种办法实现。这里重点介绍 Model 和 RedirectAttributes 携带参数的办法。

1. 重定向到页面时携带参数

（1）通过 Model 形参携带参数

在 Controller 形参中添加 Model 参数，将要传递的数据放入 Model 中进行参数传递。这种方式同样也是将参数拼接到了重定向请求的 URL 后，因而放入其中的数据只能是基本类型数据，不能是自定义类型。

项目案例：用户登录成功后，通过重定向页面实现登录后显示用户信息。（项目源码参见配套资源：第 11 章/转发与重定向/springmvc24）

关键步骤：

① 复制项目 springmvc23 为 springmvc24，修改 UserController 如下：

```
@RequestMapping("/dologin.do")
    public String doLogin(User user,Model model){
        model.addAttribute("username",user.getUsername());
        model.addAttribute("password",user.getPassword());
        return "redirect:/show.jsp";
```

}

② 测试运行，输入 http://localhost:8080/springmvc24/user/login.do。

（2）通过形参 RedirectAttributes 携带参数

RedirectAttributes 专门用于携带重定向参数。它其实继承自 Model 的接口，底层仍然使用 ModelMap 实现。所以，这种携带参数的方式同样不能携带自定义对象。

项目案例：用户登录成功后，通过重定向页面实现登录后显示用户信息。（项目源码参见配套资源：第 11 章/转发与重定向/springmvc25）

关键步骤：

① 复制项目 springmvc24 为 springmvc25，修改 UserController 如下：

```
@RequestMapping("/dologin.do")
public String doLogin(User user,RedirectAttributes rd){
    rd.addAttribute("username",user.getUsername());
    rd.addAttribute("password",user.getPassword());
    return "redirect:/show.jsp";
}
```

② 要使用 RedirectAttributes 参数，还需要在 Spring MVC 的配置文件中注册 MVC 的注解驱动：
`<mvc:annotation-driven/>`

③ 测试运行，输入 http://localhost:8080/springmvc25/user/login.do。

2. 重定向到控制器时携带参数

重定向到控制器时，携带参数的方式可以使用请求 URL 后携带方式、HttpSession 携带方式、Model 形参携带方式等，下面案例学习使用 Model 形参携带参数，注意传递与接收的要点是接收方法的形参的名称要与传递方法的 model 中的 key 名称一致，可以整体接收，也可以逐个参数接收。

项目案例：用户登录成功后，通过重定向页面实现登录后显示用户信息。（项目源码参见配套资源：第 11 章/转发与重定向/springmvc26）

关键步骤：

（1）复制 springmvc25 为 springmvc26，修改控制器 UserController 如下：

```
//重定向到控制器
@RequestMapping("/dologin.do")
public String doLogin(User user,Model model){
    model.addAttribute("username",user.getUsername());
    model.addAttribute("password",user.getPassword());
    return "redirect:second.do";
}

//逐个参数接收
@RequestMapping("/second.do")
public ModelAndView doSecond(String username,String password){
    ModelAndView mv=new ModelAndView();
    mv.addObject("username",username);
    mv.addObject("password",password);
    mv.setViewName("redirect:/show.jsp");
    return mv;
}
//整体接收
@RequestMapping("/third.do")
public ModelAndView doThird(User user){
```

```
        ModelAndView mv=new ModelAndView();
        mv.addObject("username",user.getUsername());
        mv.addObject("password",user.getPassword());
        mv.setViewName("redirect:/show.jsp");
        return mv;
}
```

（2）测试运行，输入 http://localhost:8080/springmvc26/user/login.do。

11.1.7　返回 void 时的请求转发

当处理器方法返回 void 时，可以使用 HttpServletRequest 实现请求转发，既可转发到页面，也可转发到其他控制器方法。若有数据需要向目标资源传递，可将数据放入 HttpServletRequest 或 HttpSession 中，但不能将数据放到 Model、RedirectAttributes 中，因为这两者的数据都是通过拼接到处理器方法的返回值中，作为请求的一部分向下传递的。这里没有返回值，所以它们中的数据无法向下传递。（项目源码参见配套资源：第 11 章/转发与重定向/springmvc26）

关键代码如下：

```
//无返回值的情形,由servlet来实现跳转
@RequestMapping("/dologin.do")
public void doLogin(HttpServletRequest request,
        HttpServletResponse response,User user) throws IOException, ServletException{
    if(user.getUsername().equals("张三")&&user.getPassword().equals("123")){
        request.setAttribute("user",user.getUsername());
request.getRequestDispatcher("/WEB-INF/jsp/success.jsp").forward(request,response);;
    }else{
        response.sendRedirect("login.do");
    }
}
```

11.1.8　返回 void 时的重定向

当处理器方法返回 void 时，可以使用 HttpServletResponse 的 sendRedirect()方法实现重定向。若有数据需要向下一级资源传递，需要将数据放入 HttpSession 中，而不能将其放在 HttpServletRequest 中。

项目案例： 用户登录成功后，通过重定向页面实现登录后显示用户信息。（项目源码参见配套资源：第 11 章/转发与重定向/springmvc27）

关键步骤：

（1）复制项目 springmvc26 为 springmvc27，修改控制器 UserController 如下：

```
//重定向到控制器
@RequestMapping("/dologin.do")
Public void doLogin(Useruser,HttpSession session,HttpServletRequest request,Http
ServletResponse response){
    session.setAttribute("username",user.getUsername());
    session.setAttribute("password",user.getPassword());
    try {
        response.sendRedirect(request.getContextPath()+"/show3.jsp");
    } catch (IOException e) {
        e.printStackTrace();
    }
}
```

（2）在 WebContent 下添加页面 show3.jsp，代码如下：

```
<%@ page language="java" contentType="text/html; charset=utf-8"
    pageEncoding="utf-8"%>
<!DOCTYPE html PUBLIC "-//W3C//DTD HTML 4.01 Transitional//EN" "http://www.w3.org/TR/html4/loose.dtd">
<html>
<head>
<meta http-equiv="Content-Type" content="text/html; charset=utf-8">
<title>Insert title here</title>
</head>
<body>
登录用户信息<br/>
用户名:${username}<br/>
密码:${password}<br/>
</body>
</html>
```

（3）测试运行，输入 http://localhost:8080/springmvc27/user/login.do。

11.2 异常处理

发生异常时，常用的 Spring MVC 异常处理方式主要有以下三种。
（1）使用系统定义的异常处理器 SimpleMappingExceptionResolver。
（2）使用自定义异常处理器。
（3）使用异常处理注解。

11.2.1 SimpleMappingExceptionResolver 异常处理器

该方式只需要在 Spring MVC 配置文件中注册该异常处理器 Bean 即可。该 Bean 比较特殊，没有 id 属性，无须显式调用或被注入给其他<bean/>，当异常发生时会自动执行该类。

项目案例：用户注册时会发生不同形式的异常，分别捕捉到这些异常，跳转到相应的页面显示异常信息。（项目源码参见配套资源：第 11 章/异常处理/springmvc28）

关键步骤：

（1）新建包 com.lifeng.exception，包下自定义 2 个异常类，继承自 Exception 类。

NameException 类代码如下：

```
package com.lifeng.exception;
public class NameException extends Exception {
    public NameException(){
    }
    public NameException(String message){
        super(message);
    }
}
```

AgeException 类代码如下：

```
package com.lifeng.exception;
public class AgeException extends Exception {
    public AgeException(){
```

```
        }
        public AgeException(String message){
            super(message);
        }
    }
```

（2）修改 UserController，分别出现 NameEception 异常、AgeException 异常、其他异常等情形，代码如下：

```
@RequestMapping("/doregister.do")
    public ModelAndView register(String username,int age) throws Exception {
        ModelAndView mv=new ModelAndView();
        if(!username.equals("张三")){
            throw new NameException("学生姓名不正确!");
        }
        if(age>=30){
            throw new AgeException("学生年龄太大!");
        }
        if(age==20){
            throw new Exception("测试其他异常!!,跳转到了有这些信息的页面!!!");
        }
        mv.addObject("username", username);
        mv.addObject("age",age);
        mv.setViewName("forward:/show4.jsp");
        return mv;
    }
```

（3）在 springmvc.xml 中注册异常处理器，代码如下：

```xml
<!-- 注册异常处理器 -->
        <bean class="org.springframework.web.servlet.handler.SimpleMappingExceptionResolver">
            <property name="exceptionMappings">
                <props>
                    <prop key="com.lifeng.exception.NameException">forward:/nameError.jsp</prop>
                    <prop key="com.lifeng.exception.AgeException">forward:/ageError.jsp</prop>
                </props>
            </property>
            <property name="defaultErrorView" value="forward:/defaultError.jsp"/>
            <property name="exceptionAttribute" value="ex"/>
        </bean>
```

有关属性说明如下。

① ExceptionMappings：Properties 类型属性，用于指定具体的不同类型的异常所对应的异常响应页面。Key 为异常类的全限定性类名，value 则为响应页面路径（字符串），响应页面字符串默认会受视图解释器解释，若要不受视图解释器解释，则要加上 "forward:" 前缀。

② DefaultErrorView：指定默认的异常响应页面。若发生的异常不是 exceptionMappings 中指定的异常，则使用默认异常响应页面。

③ ExceptionAttribute：捕获到的异常对象，可以在异常响应页面中使用。

（4）在 WebContent 下创建异常响应页面。

ageError.jsp 代码如下：

```jsp
<%@ page language="java" contentType="text/html; charset=utf-8" pageEncoding="utf-8"%>
<!DOCTYPE html PUBLIC "-//W3C//DTD HTML 4.01 Transitional//EN" "http://www.w3.org/TR/html4/loose.dtd">
<html>
<head>
<meta http-equiv="Content-Type" content="text/html; charset=ISO-8859-1">
<title>Insert title here</title>
</head>
<body>
${ex.message }
</body>
</html>
```

nameError.jsp 代码如下：

```jsp
<%@ page language="java" contentType="text/html; charset=utf-8"
    pageEncoding="utf-8"%>
<!DOCTYPE html PUBLIC "-//W3C//DTD HTML 4.01 Transitional//EN" "http://www.w3.org/TR/html4/loose.dtd">
<html>
<head>
<meta http-equiv="Content-Type" content="text/html; charset=ISO-8859-1">
<title>Insert title here</title>
</head>
<body>
${ex.message }
</body>
</html>
```

defaultError.jsp 代码如下：

```jsp
<%@ page language="java" contentType="text/html; charset=utf-8"
    pageEncoding="utf-8"%>
<!DOCTYPE html PUBLIC "-//W3C//DTD HTML 4.01 Transitional//EN" "http://www.w3.org/TR/html4/loose.dtd">
<html>
<head>
<meta http-equiv="Content-Type" content="text/html; charset=ISO-8859-1">
<title>Insert title here</title>
</head>
<body>
${ex.message }
</body>
</html>
```

（5）修改 WebContent/WEB-INF/jsp 下的 register.jsp 页面，代码如下：

```jsp
<%@ page language="java" contentType="text/html; charset=utf-8"
    pageEncoding="utf-8"%>
<!DOCTYPE html PUBLIC "-//W3C//DTD HTML 4.01 Transitional//EN" "http://www.w3.org/TR/html4/loose.dtd">
<html>
<head>
<meta http-equiv="Content-Type" content="text/html; charset=ISO-8859-1">
<title>Insert title here</title>
</head>
<body>
用户注册
<form action="doregister.do">
```

```
姓名:<input type="text" name="username"/><br/>
年龄:<input type="text" name="age"/><br/>
<input type="submit" value="注册"/>
</form>
</body>
</html>
```

（6）运行测试，在浏览器的地址栏中输入 http://localhost:8080/springmvc28/user/register.do，进行三次测试。

第一次测试输入和结果如图 11.3 和图 11.4 所示。

 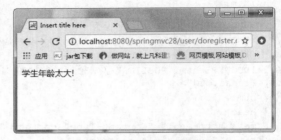

　　图 11.3　第一次测试输入　　　　　　　　　图 11.4　第一次测试结果

第二次测试输入和结果如图 11.5 和图 11.6 所示。

　　图 11.5　第二次测试输入　　　　　　　　　图 11.6　第二次测试结果

第三次测试输入和结果如图 11.7 和图 11.8 所示。

　　图 11.7　第三次测试输入　　　　　　　　　图 11.8　第三次测试结果

11.2.2　HandlerExceptionResolver 接口处理异常

上面案例使用 Spring MVC 的 SimpleMappingExceptionResolver 异常处理器，可以实现发生指定异常后的跳转。但如果想在捕获到指定异常时，执行一些操作，它就无法实现了。

解决方法之一就是让控制器实现 HandlerExceptionResolver 接口，然后重写接口的 resolveException 方法。当一个类实现了 HandlerExceptionResolver 接口后，只要有异常发生，无论什么异常，都会自动执行接口方法 resolveException()。可以通过 instanceof 来判断具体的异常种类，从而可以精确地执行一些对应的操作并跳转到不同的目标异常响应页面。实现 HandlerExceptionResolver 接口的控制器也叫自定义一个异常处理器。

项目案例：用户注册时会发生不同形式的异常，分别捕捉到这些异常，跳转到相应的页面显示异常信息。（项目源码参见配套资源：第 11 章/异常处理/springmvc29）

关键步骤：

（1）复制项目 springmvc28 为 springmvc29，删除配置文件中的 SimpleMappingExceptionResolver 异常处理器，修改控制器 UserController，让它实现 HandlerExceptionResolver 接口，如下所示：

```
public class UserController implements HandlerExceptionResolver{
```

在 UserController 中添加下述方法（原有方法不变），如下所示：

```
@Override
public ModelAndView resolveException(HttpServletRequest request,
        HttpServletResponse response, Object handler,Exception ex) {
    ModelAndView mv=new ModelAndView();
    mv.addObject("ex",ex);
    mv.setViewName("forward:/defaultError.jsp");
    if(ex instanceof NameException){
        mv.setViewName("/nameError.jsp");
    }
    if(ex instanceof AgeException){
        mv.setViewName("/ageError.jsp");
    }
    return mv;
}
```

（2）运行测试，输入 http://localhost:8080/springmvc29/user/register.do，结果同上。

11.2.3 使用@ExceptionHandler 注解实现异常处理

使用@ExceptionHandler 注解到方法上，将一个方法指定为异常处理方法。该注解只有一个可选属性 value，用于指定该注解的方法所要处理的异常类，即所要匹配的异常。被注解的方法，其返回值可以是 ModelAndView、String 或 void，方法参数可以是 Exception 及其子类对象、HttpServletRequest、HttpServletResponse 等。系统会自动为这些方法参数赋值。可以直接将异常处理方法注解于 Controller 之中。

项目案例：用户注册时会发生不同形式的异常，分别捕捉到这些异常，跳转到相应的页面显示异常信息。（项目源码参见配套资源：第 11 章/异常处理/springmvc30）

关键步骤：

（1）复制项目 springmvc29 为 springmvc30，修改 UserController，不再实现上述接口，删除 resolveException 方法。添加两个方法如下：

```
@RequestMapping("/doregister.do")
    public ModelAndView register(String username,int age) throws Exception {
        ModelAndView mv=new ModelAndView();
        if(!username.equals("张三")){
```

```java
            throw new NameException("学生姓名不正确!");
        }
        if(age>=30){
            throw new AgeException("学生年龄太大!");
        }
        if(age==20){
            throw new Exception("测试其他异常!!,跳转到了有这些信息的页面!!!");
        }
        mv.addObject("username", username);
        mv.addObject("age",age);
        mv.setViewName("forward:/show4.jsp");
        return mv;
    }
    @ExceptionHandler(NameException.class)
    public ModelAndView HandleNameException(Exception ex) {
        ModelAndView mv=new ModelAndView();
        mv.addObject("ex",ex);
        mv.setViewName("forward:/nameError.jsp");
        return mv;
    }
    @ExceptionHandler(AgeException.class)
    public ModelAndView HandleAgeException(Exception ex) {
        ModelAndView mv=new ModelAndView();
        mv.addObject("ex",ex);
        mv.setViewName("forward:/ageError.jsp");
        return mv;
    }
```

（如果匹配则转到）

这样做的话，一个控制器方法发生异常，在同一个控制器里面都可找到相匹配的异常处理方法进行处理。缺点就是正常业务代码跟异常处理代码放在同一个类，混在一起，程序的可维护性差。

（2）运行测试，输入 http://localhost:8080/springmvc30/user/register.do，结果同上。

刚才我们提到这种业务逻辑代码与异常处理代码混合编程可维护性差，解决思路是将异常处理方法专门定义在一个独立的类中，让其他 Controller 继承该类即可。不过也可能会导致新问题，因为 Java 类是"单继承"的，唯一的一个继承机会使用了，若再有其他类需要继承，将无法直接实现。

项目案例：用户注册时会发生不同形式的异常，分别捕捉到这些异常，跳转到相应的页面显示异常信息，并将异常处理方法独立成一个类。（项目源码参见配套资源：第 11 章/异常处理/springmvc31）

关键步骤：

（1）复制项目 springmvc30 为 springmvc31，在包 com.lifeng.exception 下创建专门处理异常的类 UserExceptionResolver，代码如下：

```java
package com.lifeng.exception;
import javax.servlet.http.HttpServletRequest;
import javax.servlet.http.HttpServletResponse;
import org.springframework.beans.TypeMismatchException;
import org.springframework.web.bind.annotation.ExceptionHandler;
import org.springframework.web.servlet.HandlerExceptionResolver;
import org.springframework.web.servlet.ModelAndView;
public class UserExceptionResolver{
```

```java
        //姓名异常
        @ExceptionHandler(NameException.class)
        public ModelAndView HandleNameException(Exception ex) {
            ModelAndView mv=new ModelAndView();
            mv.addObject("ex",ex);
            mv.setViewName("forward:/nameError.jsp");
            return mv;
        }
        //年龄异常
        @ExceptionHandler(AgeException.class)
        public ModelAndView HandleAgeException(Exception ex) {
            ModelAndView mv=new ModelAndView();
            mv.addObject("ex",ex);
            mv.setViewName("forward:/ageError.jsp");
            return mv;
        }
        //数据类型不匹配异常
        @ExceptionHandler(TypeMismatchException.class)
        public ModelAndView HandleMismatchException(Exception ex,HttpServletRequest request) {
            ModelAndView mv=new ModelAndView();
            String username=request.getParameter("username");
            String age=request.getParameter("age");
            mv.addObject("ex",ex);
            mv.addObject("username",username);
            mv.addObject("age",age);
            mv.setViewName("register");
            return mv;
        }
        //其他异常
        @ExceptionHandler
        public ModelAndView HandleException(Exception ex) {
            ModelAndView mv=new ModelAndView();
            mv.addObject("ex",ex);
            mv.setViewName("forward:/defaultError.jsp");
            return mv;
        }
}
```

（2）修改控制器 UserController，继承上面的异常处理类即可，代码如下：

```java
@Controller
@RequestMapping("/test")
public class UserController extends UserExceptionResolver {
    @RequestMapping("/doregister.do")
    public ModelAndView register(String username,int age) throws Exception {
        ModelAndView mv=new ModelAndView();
        if(!username.equals("张三")){
            throw new NameException("学生姓名不正确!");
        }
        if(age>=30){
            throw new AgeException("学生年龄太大!");
        }
        if(age==20){
            throw new Exception("测试其他异常!!,跳转到了有这些信息的页面!!!");
        }
```

```
                mv.addObject("username", username);
                mv.addObject("age",age);
                mv.setViewName("forward:/show4.jsp");
                return mv;
        } //省略其他方法
}
```

（3）修改 register 页面，代码如下：

```
<body>
<h2>用户注册</h2>
${ex.message }
<form action="doregister.do" method="post">
姓名:<input type="text" name="username" value="${username}"/></br>
年龄:<input type="text" name="age" value="${age}"/></br></br>
<input type="submit" value="注册"/>
</form>
</body>
```

（4）运行测试，输入 http://localhost:8080/springmvc31/user/register.do，部分测试与上面相同。下面是测试类型不匹配的情况，如图 11.9 和图 11.10 所示。

图 11.9　年龄输错了，类型不匹配

图 11.10　捕捉到异常并输出异常信息

捕捉到了类型不匹配异常，并在前台输出异常信息。此外还实现了原来填写的数据的回显，这个技术我们后面还会介绍。

当然现在这样还不够完美，因为异常信息一大堆，有点乱，后面会解决这个问题。

11.3　类型转换器

在前面的案例中，表单提交的不论是 int，还是 String 类型的请求参数，接受请求的处理器方法的形参，均可直接接收到相应类型的相应数据，无须先接收 String，再由程序员来转换，相当方便。这是因为在 Spring MVC 框架中有默认的类型转换器。这些默认的类型转换器大多数情况下可以将 String 类型的数据自动转换为相应类型的数据。

但默认类型转换器不可以将用户提交的所有 String 都转换为所有用户需要的类型。例如，Spring MVC 的默认类型转换器中，默认只能转换 "yyyy/MM/dd" 的日期格式，其他格式就无法转换，这时需要自定义类型转换器。

11.3.1 自定义类型转换器 Converter

若要定义类型转换器类，则该类需要实现 Converter 接口。该 Converter 接口有两个泛型：第一个为待转换的类型，第二个为目标类型，该接口的方法 convert()用于完成类型转换。

项目案例： 用户输入 yyyy-MM-dd 的日期字符串能被后台正确接收到。（项目源码参见配套资源：第 11 章/自定义类型转换器/springmvc32）

实现步骤：

（1）复制项目 springmvc31 为 springmvc32，修改前台注册页面 register.jsp，代码如下：

```
<body>
<h2>用户注册</h2>
${ex.message }
<form action="doregister.do" method="post">
姓名:<input type="text" name="username" value="${username}"/></br>
密码:<input type="text" name="password" value="${password}"/></br>
出生日期:<input type="text" name="birthday" value="${birthday}"/></br></br>
<input type="submit" value="注册"/>
</form>
</body>
```

（2）修改后台控制器 UserController 的 doRegister 方法，代码如下：

```
@RequestMapping("/doregister.do")
public ModelAndView doRegister(String username,String password,Date birthday) throws NameException, AgeException{
    ModelAndView mv=new ModelAndView();
    mv.addObject("username",username);
    mv.addObject("password",password);
    mv.addObject("birthday",birthday.toLocaleString());
    mv.setViewName("forward:/show4.jsp");
    System.out.println(birthday);
    return mv;
}
```

（3）新建包 com.lifeng.util 放置工具类，包下新建一个自定义日期转换类 DateConverter1，实现接口 Converter<String,Date>，代码如下：

```
public class DateConverter1 implements Converter<String,Date>{
    @Override
    public Date convert(String source) {
        try {
            if(source!=null&&!source.equals("")){
                SimpleDateFormat sdf=new SimpleDateFormat("yyyy-MM-dd");
                return sdf.parse(source);
            }
        } catch (ParseException e) {
            e.printStackTrace();
        }
        return null;
    }
}
```

（4）在 Springmvc.xml 配置文件中注册类型转换器，代码如下：

```
<bean id="dateConverter1" class="com.lifeng.controller.DateConverter1"></bean>
```

（5）在 Springmvc.xml 配置文件中创建转换服务 Bean，代码如下：
```xml
<bean id="conversionService" class="org.springframework.context.support.ConversionServiceFactoryBean">
    <property name="converters" ref="dateConverter1"></property>
</bean>
```
上面格式只能使用一种转换器，如要使用多种转换器，可以使用下面的格式：
```xml
<bean id="conversionService" class="org.springframework.context.support.ConversionServiceFactoryBean">
    <property name="converters">
        <set>
            <ref bean="dateConverter1"/>
            <ref bean="dateConverter2"/>
            <ref bean="dateConverter3"/>
        </set>
    </property>
</bean>
```
（6）Springmvc.xml 配置文件，使用类型转换服务 Bean，代码如下：
```xml
<mvc:annotation-driven conversion-service="conversionService" />
```
Springmvc.xml 最终完整的配置如下：
```xml
<?xml version="1.0" encoding="UTF-8"?>
<beans xmlns="http://www.springframework.org/schema/beans"
    xmlns:xsi="http://www.w3.org/2001/XMLSchema-instance"
    xmlns:mvc="http://www.springframework.org/schema/mvc"
    xmlns:context="http://www.springframework.org/schema/context"
    xmlns:aop="http://www.springframework.org/schema/aop"
    xmlns:tx="http://www.springframework.org/schema/tx"
    xsi:schemaLocation="http://www.springframework.org/schema/beans
        http://www.springframework.org/schema/beans/spring-beans.xsd
        http://www.springframework.org/schema/mvc
        http://www.springframework.org/schema/mvc/spring-mvc.xsd
        http://www.springframework.org/schema/context
        http://www.springframework.org/schema/context/spring-context.xsd
        http://www.springframework.org/schema/aop
        http://www.springframework.org/schema/aop/spring-aop.xsd
        http://www.springframework.org/schema/tx
        http://www.springframework.org/schema/tx/spring-tx.xsd">
    <!-- 配置视图解析器 -->
    <bean class="org.springframework.web.servlet.view.InternalResourceViewResolver">
        <!--逻辑视图前缀-->
        <property name="prefix" value="/WEB-INF/jsp/"></property>
        <!--逻辑视图后缀，匹配模式：前缀+逻辑视图+后缀，形成完整路径名-->
        <property name="suffix" value=".jsp"></property>
    </bean>
    <!-- 配置组件扫描器 -->
    <context:component-scan base-package="com.lifeng.controller"/>
    <!-- 注册类型转换器 -->
    <bean id="dateConverter1" class="com.lifeng.util.DateConverter1"></bean>
    <!-- 创建类型转换服务 Bean -->
    <bean id="conversionService" class="org.springframework.context.support.ConversionServiceFactoryBean">
        <property name="converters" ref="dateConverter1"></property>
    </bean>
```

```xml
<!-- 如果有多个类型转换器,可采用如下格式创建类型转换服务 Bean -->
<!--<bean id="conversionService" class="org.springframework.context.support.
ConversionServiceFactoryBean">
    <property name="converters">
        <set>
            <ref bean="dateConverter1"/>
            <ref bean="dateConverter2"/>
            <ref bean="dateConverter3"/>
        </set>
    </property>
</bean> -->
<!-- 注册注解驱动,使用类型转换服务 -->
<mvc:annotation-driven conversion-service="conversionService" />

</beans>
```

（7）修改 show4.jsp 如下：
```
<body>
<h2>用户信息</h2>
用户名:${username}<br/>
密码:${password}<br/>
出生日期:${birthday}<br/>
</body>
```

（8）运行测试，输入 http://localhost:8080/springmvc32/user/register.do。效果如图 11.11 和图 11.12 所示。

图 11.11　输入日期

图 11.12　正确接收到日期并输出

11.3.2　接收多种格式的日期类型转换

项目案例：无论用户输入 yyyy/MM/dd、yyyy-MM-dd，还是 yyyyMMdd，都能正确接收到日期。
（项目源码参见配套资源：第 11 章/接收多种格式的日期类型转换/springmvc33）

关键步骤：

（1）复制项目 springmvc32 为 springmvc33，修改转换器 DateConverter1 如下：
```java
@Override
    public Date convert(String source) {
        try {
            if(source!=null&&!source.equals("")){
                SimpleDateFormat sdf= getSimpleDateFormat(source);
                return sdf.parse(source);
            }
        } catch (ParseException e) {
```

```
                e.printStackTrace();
            }
            return null;
        }
        private SimpleDateFormat getSimpleDateFormat(String source){
            SimpleDateFormat sdf=new SimpleDateFormat();
            if(source.matches("^\\d{4}-\\d{2}-\\d{2}$")){
                sdf=new SimpleDateFormat("yyyy-MM-dd");
            }else if(source.matches("^\\d{4}/\\d{2}/\\d{2}$")){
                sdf=new SimpleDateFormat("yyyy/MM/dd");
            }else if(source.matches("^\\d{4}\\d{2}\\d{2}$")){
                sdf=new SimpleDateFormat("yyyyMMdd");
            }
            return sdf;

        }
```
利用正则表达式来判断日期格式。

（2）运行测试，输入 http://localhost:8080/springmvc33/user/register.do。

测试 yyyy/MM/dd 格式，效果如图 11.13 和图 11.14 所示。

图 11.13 输入日期 1

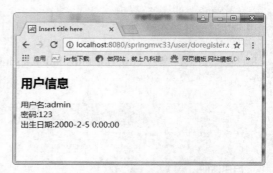
图 11.14 正确接收到日期并输出 1

测试 yyyyMMdd 格式，效果如图 11.15 和图 11.16 所示。

图 11.15 输入日期 2

图 11.16 正确接收到日期并输出 2

11.3.3 类型转换发生异常后的数据回显

当数据类型转换发生异常后，通常需要返回到表单页面，让用户重新填写。如果不处理，发生类型转换异常后，系统会自动跳转到 400 页面。若要在发生类型转换异常后跳转到指定页面，则需要捕获异常，然后通过异常处理器跳转到指定页面。仅仅完成跳转，则使用系统定义好的

SimpleMappingExceptionResolver 就可以。但如果当页面返回到原来填写的表单页面后，还需要将用户原来填写的数据显示出来，让用户能够更正填错的数据，也就是还需要完成数据回显功能，就需要自定义异常处理器了。

数据回显的原理：在异常处理器中，用 request.getParameter()方法将用户输入的表单原始数据获取到，再放入 ModelAndView 中的 Model 中，然后在要跳转到的目标页面中直接通过 EL 表达式读取出，这就实现了数据回显。

项目案例：回显类型转换失败的数据。（项目源码参见配套资源：第 11 章/类型转换失败后数据回显/springmvc34）

关键步骤：

（1）复制项目 springmvc33 为 springmvc34，修改处理器 doRegister 方法。

类型转换异常为 TypeMismatchException，代码如下：

```java
@RequestMapping("/doregister.do")
    public ModelAndView doRegister(String username,String password,int age,Date birthday) throws NameException, AgeException{
        ModelAndView mv=new ModelAndView();
        mv.addObject("username", username);
        mv.addObject("password", password);
        mv.addObject("age",age);
        mv.addObject("birthday",birthday);
        mv.setViewName("forward:/show4.jsp");
        System.out.println(birthday);
        return mv;

    }
```

（2）修改异常处理类 UserExceptionResolver 的 HandleMismatchException 方法，代码如下：

```java
//数据类型不匹配异常
@ExceptionHandler(TypeMismatchException.class)
public ModelAndView HandleMismatchException(Exception ex,HttpServletRequest request) {
    ModelAndView mv=new ModelAndView();
    String username=request.getParameter("username");
    String password=request.getParameter("password");
    String age=request.getParameter("age");
    String birthday=request.getParameter("birthday");
    mv.addObject("ex",ex);
    mv.addObject("username", username);
    mv.addObject("password", password);
    mv.addObject("age",age);
    mv.addObject("birthday",birthday);
    mv.setViewName("register");
    return mv;

}
```

（3）修改注册页面 register.jsp，代码如下：

```jsp
<body>
<h2>用户注册</h2>
${ex.message }
<form action="doregister.do" method="post">
姓名:<input type="text" name="username" value="${username}"/></br>
```

```
密码:<input type="text" name="password" value="${password}"/></br>
年龄:<input type="text" name="age" value="${age}"/></br>
注册日期:<input type="text" name="registerDate" value="${registerDate}"/></br></br>
<input type="submit" value="注册"/>
</form>

</body>
```

（4）修改 show4.jsp，代码如下：

```
<body>
<h2>用户信息</h2>
用户名:${username}<br/>
密码:${password}<br/>
年龄:${age}<br/>
注册日期:${registerDate}<br/>
</body>
```

（5）运行测试，输入 http://localhost:8080/springmvc34/user/register.do，如图 11.17～图 11.20 所示。

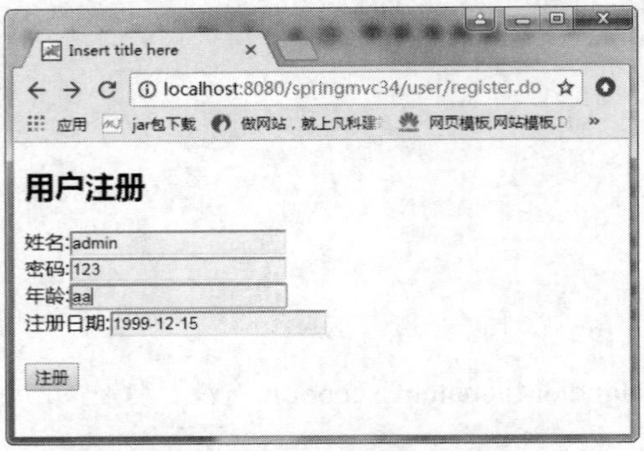

图 11.17　年龄输入异常

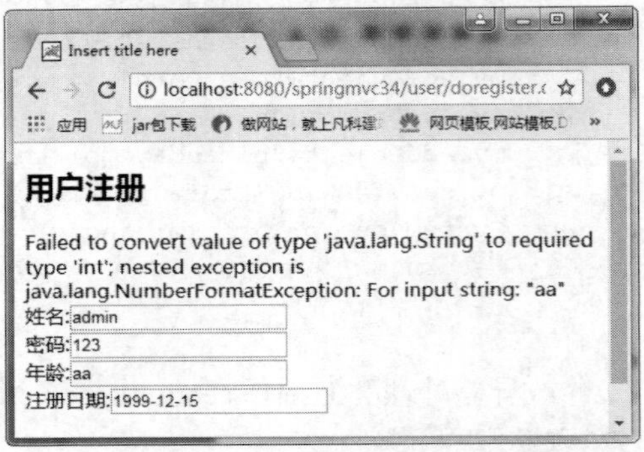

图 11.18　年龄输入异常被捕捉到并回显数据

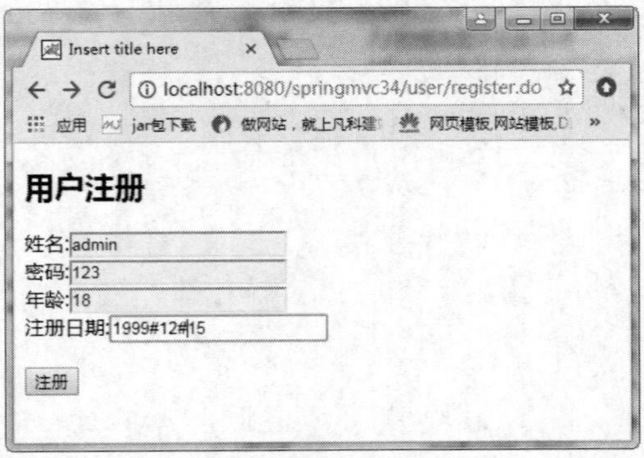

图 11.19　日期输入异常

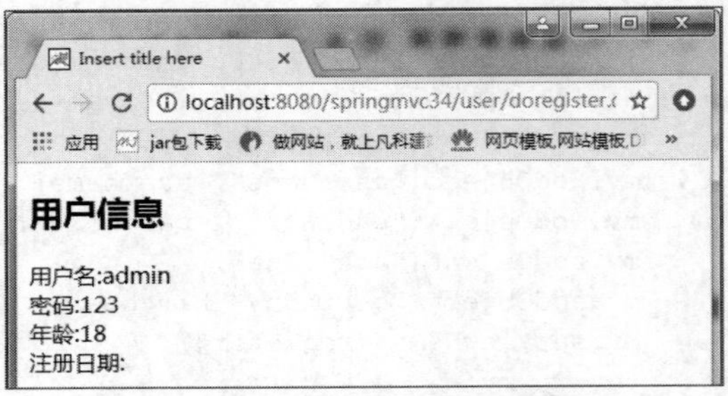

图 11.20　日期输入异常未能实现数据回显

以上代码能解决年龄输入有误，但日期输入正确的回显问题。若日期输入有误，年龄输入正确，则无法实现回现。因为当日期格式输入有误时，SimpleDateFormat 的 parse()方法将会抛出 ParseException 异常，而非 TypeMismatchException，如何解决？可否让异常处理器再捕获 ParseException，再跳转？不行，因为 Converter<S,T>的 convter()方法是没有抛出异常的，自定义类型转换器中的 convert()方法不能抛出异常，对异常的处理方式必须为 try-catch。也就是说，convert()

方法要将异常自己处理掉，而不可以抛给调用者 JVM，所以 JVM 不知道有 ParseException 异常的发生，这样异常处理器也就无法捕获到 ParseException 了。换一种思路来解决这个问题：当用户输入的日期格式不符合要求时，手工抛出一个类型匹配异常。

（6）修改类型转换器 DateConverter1 中的 getSimpleDateFormat 方法，代码如下：

```java
private SimpleDateFormat getSimpleDateFormat(String source){
    SimpleDateFormat sdf=new SimpleDateFormat();
    if(source.matches("^\\d{4}-\\d{2}-\\d{2}$")){
        sdf=new SimpleDateFormat("yyyy-MM-dd");
    }else if(source.matches("^\\d{4}/\\d{2}/\\d{2}$")){
        sdf=new SimpleDateFormat("yyyy/MM/dd");
    }else if(source.matches("^\\d{4}\\d{2}\\d{2}$")){
        sdf=new SimpleDateFormat("yyyyMMdd");
    }else{
        throw new TypeMismatchException("",Date.class);
    }
    return sdf;
}
```

手工抛出类型不匹配异常

在多重 if…else 的最后添加一个手工抛出类型匹配异常。

TypeMismatchException 类没有无参构造器，这里使用了一个带两个参数的构造器。第一个参数为要匹配的值，第二个参数为要匹配的类型，即当第一个参数的类型与第二参数的类型没有 is-a 关系时，该异常发生。

由于本例中要判断的值确定为 String 类型，而目标类型为 Date 类型，所以第一个参数就任意写了一个 String 常量。这里的判断与具体的数值无关，只与类型有关。

再次测试，如图 11.21 和图 11.22 所示。

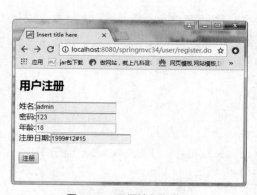

图 11.21 日期输入异常

图 11.22 日期输入异常捕捉到并数据回显成功

11.3.4 简化类型转换发生异常后的提示信息

上面的案例虽然捕捉到异常并回显了异常信息，但异常信息太烦琐，怎样才能简化这些信息呢？Spring MVC 没有专门的用于自定义类型转换失败后提示信息的功能，只能程序员自行实现。

类型转换失败后系统给出的提示信息，内容又多又乱，但仔细观察可以发现其中包含着用户所提交的输入数据。所以，可以通过在系统提示信息中查找用户输入数据的方式，来确定是哪个数据出现了类型转换异常。可以使用 String 类的 contains()方法来判断系统异常信息中是否存在用户输入的数据。一旦出问题的数据被确定，就可以用简化的信息来替换杂乱的异常信息，从而达到简化的目的。

另外替换后的异常信息也不是简单地放在头部，而是放在对应的表单的右边，以方便观察哪个表单出现了异常。

项目案例：实现转换异常提示信息的简化。（项目源码参见配套资源：第 11 章/简化异常提示信息/springmvc35）

关键步骤：

（1）复制项目 springmvc34 为 springmvc35，修改异常处理类 UserExceptionResolver，新定义一个方法，用于根据发生类型转换异常的数据的不同而生成不同简化后的提示信息，代码如下：

```
//用来替换异常信息
public String replaceExceptionMessage(Exception ex,String data,String paramName){
    String exMsg=ex.getMessage();
    if(exMsg.contains(data)){
        String newMsg=paramName+"["+data+"]"+"填写有误!";
        return newMsg;
    }
    return null;
}
```

之后修改异常处理方法，在异常处理方法中将新生成的异常信息写入 ModelAndView 中的 Model，用于在页面中通过 EL 表达式显示，代码如下：

```
//数据类型不匹配异常
@ExceptionHandler(TypeMismatchException.class)
public ModelAndView HandleMismatchException(Exception ex,HttpServletRequest request) {
    ModelAndView mv=new ModelAndView();
    String username=request.getParameter("username");
    String password=request.getParameter("password");
    String age=request.getParameter("age");
    String registerDate=request.getParameter("registerDate");
    mv.addObject("username", username);
    mv.addObject("password", password);
    String ageMsg=replaceExceptionMessage(ex,age,"年龄");
    mv.addObject("age",age);
    mv.addObject("ageMsg",ageMsg);
    String registerDateMsg=replaceExceptionMessage(ex,registerDate,"注册日期");
    mv.addObject("registerDate",registerDate);
    mv.addObject("registerDateMsg",registerDateMsg);
    mv.setViewName("register");
    return mv;
}
```

（2）修改前台页面 register.jsp，代码如下：

```
<form action="test/register.do" method="post">
姓名:<input type="text" name="name" /></br>
年龄:<input type="text" name="age" value="${age}"/>${ageMsg}</br>
```

```
注册日期:<input type="text" name="registerDate" value="${registerDate}"/></br></br>
<input type="submit" value="注册"/>
</form>
```

（3）测试结果如图 11.23 和图 11.24 所示。

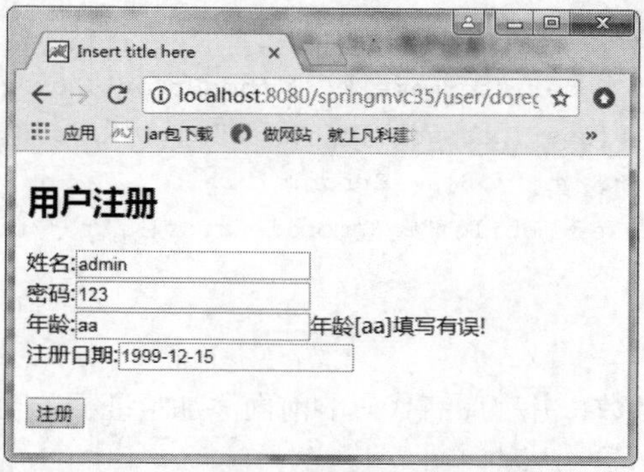

图 11.23　年龄输入异常捕捉到并简化提示异常信息

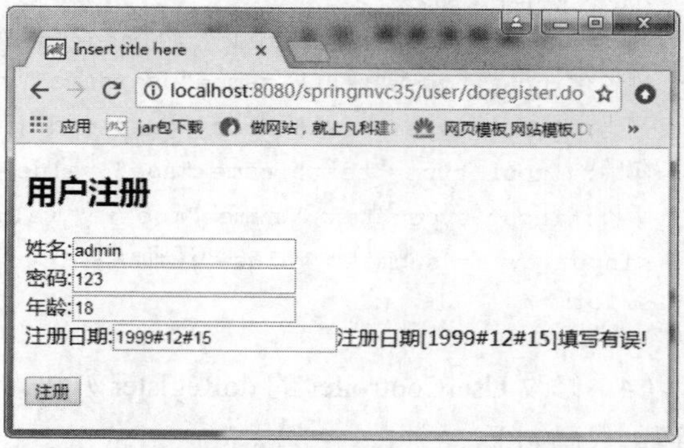

图 11.24　日期输入异常捕捉到并简化提示异常信息

11.4　数据验证

在 Web 应用程序中，为了防止客户端传来的数据导致程序的异常，常常需要对数据进行验证。验证分为客户端验证与服务器端验证两种。客户端验证主要通过 JavaScript 脚本进行，服务器端验证主要是通过 Java 代码进行。

一般情况下，客户端验证与服务器端验证都是要进行的，以保证数据的安全性，本节介绍的是 Spring MVC 服务器端数据验证。

项目案例： 对用户注册信息进行服务器端验证，要求用户输入的表单数据满足如下要求。

姓名：非空，且长度为 3~6 个字符。

密码：非空，长度为 6，只能由数字或字母组成。

年龄：18~30 岁。

手机号：非空，且必须符合手机号的格式。

（项目源码参见配套资源：第 11 章/数据验证/springmvc36）

实现步骤：

（1）复制项目 springmvc35 为 springmvc36，导入验证有关的三个 JAR 包。

除了 SpringMVC 的 JAR 包外，我们还需要导入 Hibernate、Validator 的 JAR 包。这些 JAR 包可以从 Hibernate 官网中直接下载：hibernate-validator-4.3.0.final.jar、jboss-logging-3.3.0.Final.jar、validation-api-1.0.0.GA.jar。

（2）修改实体类 User 如下：

```
package com.lifeng.entity;
public class User {
    String username;
    String password;
    int age;
    String mobile;
```

        //省略getter,setter方法
}

（3）修改注册页面register.jsp如下：
```
<body>
<h2>用户注册</h2>
<form action="doregister.do" method="post">
姓名:<input type="text" name="username" value="${user.username}"/>${nameError}</br>
密码:<input type="text" name="password" value="${user.password}"/>${pwdError}</br>
年龄:<input type="text" name="age" value="${user.age}"/>${ageError}</br>
手机:<input type="text" name="mobile" value="${user.mobile}"/>${mobileError}</br></br>
<input type="submit" value="注册"/>
</form>
</body>
```

（4）修改UserController的doRegister方法，修改参数，并在原有代码的前面添加验证相关功能代码，代码如下：
```
@RequestMapping("/doregister.do")
    public ModelAndView doRegister(@Validated User user,BindingResult br){
            //先进行数据验证,如果验证不通过跳回注册页面register.jsp
            List<ObjectError> errors=br.getAllErrors();
            if(errors.size()>0){
                ModelAndView m1=new ModelAndView();
                FieldError nameError=br.getFieldError("username");
                FieldError pwdError=br.getFieldError("password");
                FieldError ageError=br.getFieldError("age");
                FieldError mobileError=br.getFieldError("mobile");
                if(nameError!=null){
                    m1.addObject("nameError",nameError.getDefaultMessage());
                }
                if(pwdError!=null){
                    m1.addObject("pwdError",pwdError.getDefaultMessage());
                }
                if(ageError!=null){
                    m1.addObject("ageError",ageError.getDefaultMessage());
                }
                if(mobileError!=null){
                    m1.addObject("mobileError",mobileError.getDefaultMessage());
                }
                m1.addObject("user",user);
                m1.setViewName("register");
                return m1;
            }
            //原有代码,验证通过才会进行下面的操作
            ModelAndView mv=new ModelAndView();
            mv.addObject("user",user);
            mv.setViewName("forward:/show4.jsp");
            return mv;
    }
```

第一个参数说明：由于验证器为Bean对象验证器，所以要验证的参数数据需要由处理器方法以Bean形参类型的方式整体接收，并使用注解@Validated标注，但不能将@Validated注解在String类型与基本类型的形参前。

第二个参数是一个 BindingResult 类型的形参。通过这个形参可获取到所有验证异常信息，只要发生数据验证失败，就需要将页面重新跳转到 register.jsp 表单页面，让用户重填。

BindingResult 接口中常用的方法如下。

- getAllErrors()：获取到所有的异常信息。其返回值为 List，但若没有发生异常，则该 List 也会被创建，只不过其 size()为 0，而非 List 为 Null。
- getFieldError()：获取指定属性的异常信息。
- getErrorCount()：获取所有异常的数量。
- getRawFieldValue()：获取到用户输入的引发验证异常的原始值。

（5）修改 Spring MVC 配置文件，注册验证器 Bean，并在注解驱动上使用该验证器，代码如下：

```
<!-- 注册数据验证器 Bean -->
    <bean id="validator" class="org.springframework.validation.beanvalidation.LocalValidatorFactoryBean">
        <property name="providerClass" value="org.hibernate.validator.HibernateValidator"></property>
    </bean>

    <!-- 注册注解驱动,使用类型转换服务,同时使用验证器 -->
    <mvc:annotation-driven conversion-service="conversionService" validator= "validator"/>
```

（6）在实体类上添加验证注解，代码如下：

```
package com.lifeng.entity;
import javax.validation.constraints.Max;
import javax.validation.constraints.Min;
import javax.validation.constraints.Pattern;
import javax.validation.constraints.Size;
import org.hibernate.validator.constraints.Length;
import org.hibernate.validator.constraints.NotEmpty;
public class User {
    @NotEmpty(message="姓名不能为空!")
    @Length(min=3,max=6,message="长度必须在{min}与{max}之间")
    String username;

    @NotEmpty(message="姓名不能为空!")
    @Length(min=6,max=6,message="长度必须为{min}")
    String password;

    @Min(value=18,message="年龄不能小于{value}")
    @Max(value=30,message="年龄不能大于{value}")
    int age;

    @NotEmpty(message="手机号不能为空!")
    @Pattern(regexp="^1[34578]\\d{9}$",message="手机号码格式不正确")
    String mobile;
    //省略 getter,setter 方法
}
```

数据验证中可能用到的一些校验规则，如表 11.1 所示。

表 11.1 常用的数据验证规则

注解	验证规则
@AssertFalse	验证注解的元素必须为 false
@AssertTrue	验证注解的元素必须为 true
@DecimalMax(value)	验证注解的元素必须为一个数字，其值必须小于等于指定的最小值
@DecimalMin(Value)	验证注解的元素必须为一个数字，其值必须大于等于指定的最小值
@Digits(integer=, fraction=)	验证注解的元素必须为一个数字，其值必须在可接受的范围内
@Future	验证注解的元素必须是日期，检查给定的日期是否比现在晚
@Max(value)	验证注解的元素必须为一个数字，其值必须小于等于指定的最小值
@Min(value)	验证注解的元素必须为一个数字，其值必须大于等于指定的最小值
@NotNull	验证注解的元素必须不为 null
@Null	验证注解的元素必须为 null
@Past(java.util.Date/Calendar)	验证注解的元素必须是过去的日期，检查标注对象中的值表示的日期要比当前早
@Pattern(regex=, flag=)	验证注解的元素必须符合正则表达式，检查该字符串是否能够在 match 指定的情况下被 regex 定义的正则表达式匹配
@Size(min=, max=)	验证注解的元素必须在制定的范围（数据类型：String、Collection、Map and arrays）
@Valid	递归地对关联对象进行校验，如果关联对象是个集合或者数组，那么对其中的元素进行递归校验，如果是一个 map，则对其中的值部分进行校验
@CreditCardNumber	对信用卡号进行一个大致的验证
@Email	验证注释的元素必须是电子邮箱地址
@Length(min=, max=)	验证注解的对象必须是字符串，其长度在指定的范围内
@NotBlank	验证注解的对象必须是字符串，不能为空，检查时会将空格忽略
@NotEmpty	验证注释的对象必须为空（数据：String、Collection、Map、arrays）
@Range(min=, max=)	验证注释的元素必须在合适的范围内（数据：BigDecimal、BigInteger、String、byte、short、int、long 和原始类型的包装类）
@URL(protocol=, host=, port=, regexp=, flags=)	验证注解的对象必须是字符串，检查是否是一个有效的 URL，如果提供了 protocol、host 等，则该 URL 还需满足提供的条件

（7）此外 show4.jsp 也简单修改一下，如下所示：
```
<body>
<h2>用户信息</h2>
用户名:${user.username}<br/>
密码:${user.password}<br/>
年龄:${user.age}<br/>
手机:${user.mobile}<br/></body>
```
（8）运行测试，如图 11.25 和图 11.26 所示。

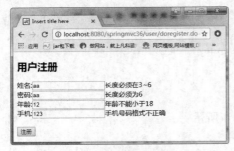

图 11.25 数据验证成功

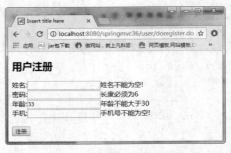

图 11.26 数据验证测试

11.5 文件上传

11.5.1 上传单个文件

项目案例：上传单个图片文件。（项目源码参见配套资源：第 11 章/上传单个图片/springmvc37）
实现步骤：
（1）复制项目 springmvc36 为 springmvc37，导入下列两个 JAR 包。
Spring MVC 实现文件上传，需要再添加两个 JAR 包。一个是文件上传的 JAR 包 commons-fileupload-1.2.2.jar，一个是其所依赖的 I/O 包 commons-io-2.4.jar。
（2）在 WebContent 下新建文件夹 images 用于存储上传的图片，为了防止空的 image 文件夹不能成功地部署到 tomcat 服务器，故先在 image 下任意添加一个图片文件，在 WebContent 下创建页面 index.jsp，表单 method 属性为 POST，enctype 属性为 multipart/form-data，代码如下：

```
<form action="user/upload.do" method="post" enctype="multipart/form-data">
照片:<input type="file" name="photo"/></br>
<input type="submit" value="上传"/>
</form>
```

（3）创建控制器方法 doUpload 实现上传功能，代码如下：

```
@RequestMapping("/upload.do")
    public ModelAndView doUpload(MultipartFile photo, HttpSession session) throws IOException{
        if(!photo.isEmpty()){
            String path=session.getServletContext().getRealPath("\\images");
            String filename=photo.getOriginalFilename();
            if(filename.endsWith(".jpg")||filename.endsWith(".png")){
                File file=new File(path,filename);
                photo.transferTo(file);
                ModelAndView mv=new ModelAndView();
                mv.addObject("filename",filename);
                mv.setViewName("forward:/success.jsp");
                return mv;
            }
        }
        return new ModelAndView("forward:/fail.jsp");
    }
```

处理器方法的定义需要注意以下几点。
① 处理器方法的形参。
注意，用于接收表单元素所提交参数的处理器方法的形参类型是 MultipartFile，而不是 File。MultipartFile 是一个专门用于处理文件上传的接口，该接口常用的方法如下。
- getName()获取参数名称。
- getOriginalFilename()获取文件的原始名称。
- getSize()获取文件大小。
- isEmpty()判断文件是否为空。
- transferTo()文件上传。

MultipartFile 接口常用的实现类为 CommonsMultipartResolver。而该实现类中具有设置上传文件大小、上传文件字符集等属性，可以在 springmvc.xml 配置文件中通过为其注入值来限定上传的文件。

② 如何判断有无上传文件。

如果用户未选择上传的文件就直接提交了表单，则这时处理器方法的 MultipateFile 形参所接收到的实参值是一个内容为 empty 的文件（注意并不是 null），所以对未选择上传文件的情况的处理，其判断条件是 file.isEmpty()，而并非 file == null。

③ 如何判断上传文件类型并限制。

Spring MVC 的文件上传功能并没有直接地用于限定文件上传类型的方法或属性，需要自行对获取到的文件名的后缀加以判断。可使用 String 的 endWith() 方法。

④ 关于上传方法 transferTo。

直接使用 MultipartFile 的 transferTo() 方法，就可以完成单个文件的上传功能。需要注意的是该方法要求服务器端用于存放客户上传文件的目录必须存在，否则报错，即其不会自己创建该目标目录。例如，本例中必须手工创建 images 目录。

（4）在 springmvc.xml 配置文件中注册文件上传处理器，代码如下：

```xml
<bean id="multipartResolver" class="org.springframework.web.multipart.commons.CommonsMultipartResolver">
    <property name="defaultEncoding" value="utf-8"/>
    <property name="maxUploadSize" value="1048576"/>

</bean>
```

对上述配置需要注意如下几点。

① Bean 名称必须是 multipartResolver。在 Spring MVC 的配置文件中注册 MultipartFile 接口的实现类 CommonsMultipartResolver 的 Bean。要求该 Bean 的 id 必须为 multipartResolver。

② 文件上传字符集解决中文文件名乱码问题。文件名为中文的文件，默认情况下上传完成后，文件名为乱码，因为默认情况下文件上传处理器使用的字符集为 ISO-8859-1。可以通过设置属性 defaultEncoding 来指定文件上传所使用的字符集。

③ 限定文件大小。MultipartFile 接口的实现类 CommonsMultipartResolver 继承自 CommonsFileUploadSupport 类，而该类有一个属性 maxUploadSize 可以用来限定上传文件的大小，单位是字节（B）。如果不对该属性进行设置，或指定其值为–1，则表示对上传文件大小无限制。还要注意，该值为上传文件的总大小，如果上传了多个文件，则多个文件的合计大小不能超过该设定值。另外还可以设置属性 maxUploadSizePerFile，来限制每个文件的大小不能超过 maxUploadSizePerFile 指定值，当然多个文件大小总和也不能超过 maxUploadSize 指定值。

当上传文件超出指定大小时，会抛出 MaxUploadSizeExceededException 异常。通过在 Spring MVC 配置文件中注册 SimpleMappingExceptionResolver 异常处理器，配置默认异常处理即可实现对该异常的处理。

（5）注册异常处理器如下代码所示，注意方框部分。

```
<!-- 注册异常处理器 -->
<bean class="org.springframework.web.servlet.handler.SimpleMappingExceptionResolver">
    <property name="exceptionMappings">
        <props>
            <prop key="com.lifeng.exception.NameException">forward:/nameError.jsp</prop>
            <prop key="com.lifeng.exception.AgeException">forward:/ageError.jsp</prop>
        </props>
    </property>
    <property name="defaultErrorView" value="forward:/defaultError.jsp"/>
    <property name="exceptionAttribute" value="ex"/>
</bean>
```

（6）创建异常响应页面，即成功与失败页面。

上传成功页面 success.jsp 代码如下：

```
<%@ page language="java" contentType="text/html; charset=utf-8"
    page Encoding="utf-8"%>
<%
String path = request.getContextPath();
String basePath = request.getScheme()+"://"+request.getServerName()+":"+request.getServerPort()+path+"/";
request.setAttribute("basePath", basePath);
%>
<!DOCTYPE html PUBLIC "-//W3C//DTD HTML 4.01 Transitional//EN" "http://www.w3.org/TR/html4/loose.dtd">
<html>
<head>
<meta http-equiv="Content-Type" content="text/html; charset=ISO-8859-1">
<title>Insert title here</title>
</head>
<body>
上传成功<br/>
<img src="${basePath}/images/${filename}" width="200" height="150"/>
</body>
</html>
```

上传失败页面 fail.jsp 代码如下：

```
<body>
上传失败!
</body>
```

（7）测试运行，输入 http://localhost:8080/springmvc37，效果如图 11.27～图 11.29 所示。

图 11.27　初始上传界面

图 11.28　选择了一个图片文件

图 11.29 上传文件成功并显示

11.5.2 上传多个文件

基本流程跟上传单个文件的流程是差不多的，只是控制器方法需要用数组接收有关参数并进行遍历。

项目案例：上传多个图片文件。（项目源码参见配套资源：第 11 章/上传多个图片/springmvc38）

实现步骤：

（1）复制项目 springmvc37 为 springmvc38，修改 index.jsp 页面如下：

```
<%@ page language="java" contentType="text/html; charset=utf-8" pageEncoding="utf-8"%>
<!DOCTYPE html PUBLIC "-//W3C//DTD HTML 4.01 Transitional//EN" "http://www.w3.org/TR/html4/loose.dtd">
<html>
<head>
<meta http-equiv="Content-Type" content="text/html; charset=utf-8">
<title>Insert title here</title>
</head>
<body>
<br/><br/>上传多个文件<br/>
<form action="user/upload2.do" method="post" enctype="multipart/form-data">
照片1:<input type="file" name="photos"/><br/>
照片2:<input type="file" name="photos"/><br/>
照片3:<input type="file" name="photos"/><br/>
<input type="submit" value="上传"/>
</form>
</body></html>
```

【注意】多个 file 表单元素名称相同。

（2）在控制器 UserController 中添加方法 upload2 如下：

```
@RequestMapping("/upload2.do")
public ModelAndView upload2(MultipartFile[] photos, HttpSession session) throws IOException{
    String path=session.getServletContext().getRealPath("/images/");
    String[] filenames=new String[photos.length];
    ModelAndView mv=new ModelAndView();
    for(int i=0;i<photos.length;i++){
```

```java
            if(!photos[i].isEmpty()){
                String filename=photos[i].getOriginalFilename();
                if(filename.endsWith(".jpg")||filename.endsWith(".png")){
                    File file=new File(path,filename);
                    photos[i].transferTo(file);
                    filenames[i]=filename;
                    mv.addObject("filenames",filenames);
                    mv.setViewName("forward:/success.jsp");
                }else{
                    return new ModelAndView("forward:/fail.jsp");
                }
            }
        }
    }
    return mv;
}
```

用于接收表单元素所提交参数的处理器方法的形参类型为 MultipartFile 数组。

（3）修改上传成功页面 success.jsp 如下：

```jsp
<%@ page language="java" contentType="text/html; charset=utf-8" pageEncoding="utf-8"%>
<%@ taglib prefix="c" uri="http://java.sun.com/jsp/jstl/core" %>
<%
String path = request.getContextPath();
String basePath=request.getScheme()+"://"+request.getServerName()+":"+request.getServerPort()+path+"/";
request.setAttribute("basePath", basePath); %>
<!DOCTYPE html PUBLIC "-//W3C//DTD HTML 4.01 Transitional//EN" "http://www.w3.org/TR/html4/loose.dtd">
<html>
<head>
<meta http-equiv="Content-Type" content="text/html; charset=ISO-8859-1">
<title>Insert title here</title>
</head>
<body>
上传多个图片成功<br/>
<c:forEach items="${filenames}" var="filename">
<img src="${basePath}/images/${filename}" width="200" height="150"/><br/><br/>
</c:forEach>
</body></html>
```

由于这里用到了 JSTL 标签，所以还要导入 jstl-1.2.jar 包。

（4）运行测试，如图 11.30 和图 11.31 所示。

图 11.30　选择多个图片文件

图 11.31 多个图片文件上传成功并显示

11.6 文件下载

项目案例：实现上传一个图片文件，然后显示出来并提供下载的超链接，单击下载。（项目源码参见配套资源：第 11 章/文件下载/springmvc39）

实现步骤：

（1）复制项目 springmvc37 为 springmvc39，在 success.jsp 添加下载的超链接：

```
<a href="download.do?filename=${filename}">下载</a>
```

（2）创建实现下载功能的控制器方法，如下所示：

```
@RequestMapping("/download.do")
    public ResponseEntity<byte[]> fileDownLoad(String filename,HttpServletRequest request) throws IOException{
        String path=request.getServletContext().getRealPath("/images/");
        File file=new File(path,filename);
        HttpHeaders headers=new HttpHeaders();
        headers.setContentDispositionFormData("attachment", filename);
        headers.setContentType(MediaType.APPLICATION_OCTET_STREAM);
        return new ResponseEntity<byte[]>(FileUtils.readFileToByteArray(file),headers,HttpStatus.OK);
    }
```

（3）测试，先上传图片文件，如图 11.32 所示。

单击"下载"，如图 11.33 所示。

图 11.32 先上传再下载

图 11.33 浏览器显示下载成功

上面是英文名称的图片，那如果是中文名称呢？如何下载中文名称的文件？

项目案例：实现上传一个中文名称的图片文件，然后显示出来并提供下载的超链接。（项目源码参见配套资源：第 11 章/文件下载/springmvc40）

实现步骤：

（1）复制 springmvc39 为 springmvc40，修改 success.jsp 下载页面，在页头添加一行代码，导入 java.net.URLEncoder 如下。

```
<%@page import="java.net.URLEncoder" %>
```

添加一个下载的超链接如下。

```
<a href="download2.do?filename=${filename}">下载中文名称的图片</a>
```

完整代码如下。

```
<%@ page language="java" contentType="text/html; charset=utf-8" pageEncoding="utf-8"%>
<%@page import="java.net.URLEncoder" %>
<%
String path = request.getContextPath();
String basePath = request.getScheme()+"://"+request.getServerName()+":"+request.getServerPort()+path+"/";
request.setAttribute("basePath", basePath);
%>
<!DOCTYPE html PUBLIC "-//W3C//DTD HTML 4.01 Transitional//EN" "http://www.w3.org/TR/html4/loose.dtd">
<html>
<head>
<meta http-equiv="Content-Type" content="text/html; charset=ISO-8859-1">
<title>Insert title here</title>
</head>
<body>
上传成功<br/>
<img src="${basePath}/images/${filename}" width="200" height="150"/><br/>
<a href="download.do?filename=${filename}">下载</a><br/>
<a href="download2.do?filename=${filename}">下载中文名称的图片</a>
</body></html>
```

（2）在 UserController 中添加处理器方法 fileDownLoad2 如下。这个方法有点复杂，通常直接复制使用即可。

```
@RequestMapping("/download2.do")
    public ResponseEntity<byte[]> fileDownLoad2(String filename,HttpServletRequest
```

```
request) throws IOException{
        String path=request.getServletContext().getRealPath("/images/");
        filename=filename.replace("_", "%");
        filename=java.net.URLDecoder.decode(filename,"utf-8");
        String downfilename="";
        if(request.getHeader("USER-AGENT").toLowerCase().indexOf("msie")>0){
            filename=URLEncoder.encode(filename,"utf-8");
            downfilename=filename.replaceAll("+", "%20");
        }else{
            downfilename=new String(filename.getBytes("UTF-8"),"ISO-8859-1");
        }
        File file=new File(path,filename);
        HttpHeaders headers=new HttpHeaders();
        headers.setContentDispositionFormData("attachment", downfilename);
        headers.setContentType(MediaType.APPLICATION_OCTET_STREAM);
        return new ResponseEntity<byte[]>(FileUtils.readFileToByteArray(file),
headers,HttpStatus.OK);
    }
```

（3）运行测试，效果如图 11.34 所示。

如果单击第一个按钮，下载的将不是中文名的图片文件，所以本案例提供了两个超链接供对比。

图 11.34　下载中文名图片文件成功

11.7　拦截器

Spring MVC 的拦截器 Interceptor 主要用来拦截指定的用户请求，并进行相应的预处理或后处理。其拦截的时间点是在处理器适配器执行处理器之前。创建拦截器类需要实现 HandlerInterceptor 接口，然后在配置文件中注册并指定拦截目标。

11.7.1　单个拦截器的执行流程

项目案例：创建第一个拦截器。（项目源码参见配套资源：第 11 章/拦截器/springmvc41）

实现步骤：

（1）复制 springmvc8 为 springmvc41，在处理器 UserController 中添加方法如下：

```
@RequestMapping("/test1.do")
public String doTestInterceptor(){
```

```
        System.out.println("执行了处理器的方法!");
        return "welcome";
    }
```
（2）新建包 com.lifeng.intercepter，在包下新建一个自定义拦截器类 Intercepter1，实现 HandlerInterceptor 接口。重写以下 3 个方法：
```
public class Intercepter1 implements HandlerInterceptor{
    @Override
    public void afterCompletion(HttpServletRequest request, HttpServletResponse response, Object handler, Exception arg3)throws Exception {
        System.out.println("执行了 Intercepter1 ----------afterCompletion");
    }
    @Override
    public void postHandle(HttpServletRequest request, HttpServletResponse response, Object handler, ModelAndView modelAndView)throws Exception {
        System.out.println("执行了 Intercepter1 ----------postHandle");
    }
    @Override
    public boolean preHandle(HttpServletRequest request, HttpServletResponse response, Object handler) throws Exception {
        System.out.println("执行了 Intercepter1 ----------preHandle");
        return true;
    }
}
```
自定义拦截器需要实现 HandlerInterceptor 接口中的 3 个方法，具体如下。

① preHandle(request,response,Object handler)：该方法在处理器方法执行之前执行。其返回值为 boolean，若为 true，则紧接着会执行处理器方法，且会将 afterCompletion()方法放入一个专门的方法栈中等待执行。若为 false，则不会执行处理器方法。

② postHandle(request,response,Object handler,modelAndView)：该方法在处理器方法执行之后执行。处理器方法若最终未被执行，则该方法不会执行。由于该方法是在处理器方法执行完后执行，且该方法参数中包含 ModelAndView，所以该方法可以修改处理器方法的处理结果数据，且可以修改跳转方向。

③ afterCompletion(request,response,Object handler,Exception ex)：当 preHandle()方法返回 true 时，会将该方法放到专门的方法栈中，等到对请求进行响应的所有工作完成之后才执行该方法。

拦截器中方法与处理器方法的执行顺序如图 11.35 所示。

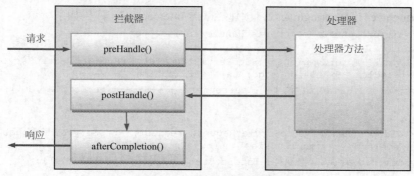

图 11.35　单个拦截器执行流程 1

也可以这样来看，如图 11.36 所示。

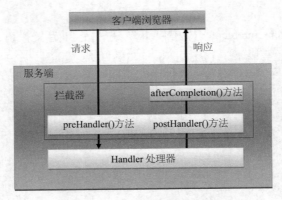

图 11.36　单个拦截器执行流程 2

（3）在 springmvc.xml 配置文件中注册拦截器，代码如下：

```
<mvc:interceptors>
    <mvc:interceptor>
        <mvc:mapping path="/**"/>
        <bean class="com.lifeng.controller.MyIntercepter"/>
    </mvc:interceptor>
    </mvc:interceptor>
</mvc:interceptors>
```

<mvc:mapping/>用于指定当前所注册的拦截器可以拦截的请求路径，/**表示拦截所有请求，/*.do 表示拦截所有名称为.do 结尾的请求。

（4）测试运行，输入 http://localhost:8080/springmvc41/user/test1.do，控制台输出如下：

```
执行了 Intercepter1 ----------preHandle
执行了处理器的方法！
执行了 Intercepter1 ----------postHandle
执行了 Intercepter1 ----------afterCompletion
```

11.7.2　多个拦截器的执行

项目案例：创建多个拦截器。（项目源码参见配套资源：第 11 章/拦截器/springmvc42）

实现步骤：

（1）复制 springmvc41 为 springmvc42，创建拦截器 Intercepter2，代码如下所示：

```
public class Intercepter2 implements HandlerInterceptor{
    @Override
    public void afterCompletion(HttpServletRequest request, HttpServletResponse response, Object handler, Exception arg3) throws Exception {
        System.out.println("执行了 Intercepter2----------afterCompletion");
    }
    @Override
    public void postHandle(HttpServletRequest request, HttpServletResponse response, Object handler, ModelAndView modelAndView) throws Exception {
        System.out.println("执行了 Intercepter2----------postHandle");
    }
    @Override
    public boolean preHandle(HttpServletRequest request, HttpServletResponse response,
```

```
Object handler) throws Exception {
        System.out.println("执行了 Intercepter2----------preHandle");
        return true;
    }
}
```
（2）注册多个拦截器，代码如下：
```
<mvc:interceptors>
    <mvc:interceptor>
        <mvc:mapping path="/*.do"/>
        <bean class="com.lifeng.controller.Intercepter1"/>
    </mvc:interceptor>

    <mvc:interceptor>
        <mvc:mapping path="/*.do"/>
        <bean class="com.lifeng.controller.Intercepter2"/>
    </mvc:interceptor>

</mvc:interceptors>
```
（3）运行测试，输入 http://localhost:8080/springmvc42/user/test1.do，结果如下：

执行了 Intercepter1 ----------preHandle
执行了 Intercepter2----------preHandle
执行了处理器的方法！
执行了 Intercepter2----------postHandle
执行了 Intercepter1 ----------postHandle
执行了 Intercepter2----------afterCompletion
执行了 Intercepter1 ----------afterCompletion

当有多个拦截器时，形成拦截器链。拦截器链的执行顺序与其注册顺序一致。需要再次强调一点的是当某一个拦截器的 preHandle()方法返回 true 并被执行到时，会向一个专门的方法栈中放入该拦截器的 afterCompletion()方法。

多个拦截器方法与处理器方法的执行顺序如图 11.37 所示。

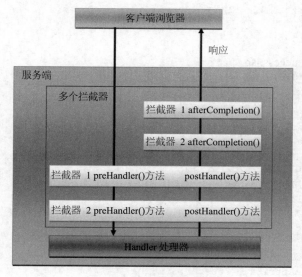

图 11.37　多个拦截器执行流程

只要有一个 preHandle()方法返回 false，则上部的执行链将被断开，其后续的处理器方法与 postHandle()方法将无法执行。但无论执行链执行情况怎样，只要方法栈中有方法，即执行链中只要有 preHandle()方法返回 true，就会执行方法栈中的 afterCompletion()方法，最终都会给出响应。

11.7.3 权限拦截器

项目案例：只有经过登录的用户方可访问处理器，否则，将返回"无权访问"提示。本例的登录由一个 JSP 页面完成，即在该页面里将用户信息放入 session 中。也就是说，只要访问过该页面，就说明登录了，没访问过，则为未登录用户。（项目源码参见配套资源：第 11 章/拦截器/springmvc43）

实现步骤：

（1）复制 springmvc42 为 springmvc43，修改处理器 UserController，添加方法 doAdmin，一个登录用户可以访问的处理方法：

```
@RequestMapping("/admin.do")
public String doAdmin(){
    System.out.println("欢迎你进入系统!");
    return "admin";
}
```

（2）在 WebContent/WEB-INF/jsp 下创建 admin.jsp 页面，代码如下：

```
<body>
<h2>这是授权用户才能进入的页面</h2>
</body>
```

（3）登录页面 login.jsp 如下：

```
<body>
<h2>用户登录</h2>
<font color="red">${msg }</font><br/>
<form action="dologin.do">
姓名:<input type="text" name="username"/><br/>
密码:<input type="text" name="password"/><br/><br/>
<input type="submit" value="登录"/>
</form>
</body>
```

（4）登录成功页面 success.jsp 如下：

```
<body>
<h2>登录成功,欢迎${user.username}光临</h2>
<a href="logout.do">退出登录</a>
<a href="admin.do">执行授权操作</a>
</body>
```

（5）控制器 UserController 有关方法如下：

```
@RequestMapping("/login.do")
public String login(){
    return "login";
}
@RequestMapping("/logout.do")
public String logout(HttpSession session){
    session.removeAttribute("user");
    return "login";
```

```
}
@RequestMapping("/dologin.do")
public String doLogin(User user,HttpSession session){
    if(user.getUsername().equals("admin")
            &&user.getPassword().equals("123")){
        session.setAttribute("user","admin");
        return "success";
    }
    return "login";
}
@RequestMapping("/admin.do")
public String admin(){
    System.out.println("欢迎你进入系统!");
    return "admin";
}
```

（6）测试正常登录。访问 http://localhost:8080/user/login.do，如图 11.38 所示。

图 11.38　登录成功界面

通过超链接访问授权用户操作，如图 11.39 所示。

图 11.39　授权用户页面

退出登录，直接访问授权操作，在浏览器中直接输入 http://localhost:8080/user/admin.do 也能访问到，可见这个系统存在安全漏洞，未登录的用户也能访问到授权用户专用的操作。解决问题的思路就是利用拦截器，拦截一切请求，先判断是否为授权用户，是授权用户才继续下一步的操作，不是则返回。

（7）在包 com.lifeng.intercepter 下创建权限拦截器 PermissionIntercepter，代码如下：

```
public class PermissionIntercepter implements HandlerInterceptor{
    @Override
    public boolean preHandle(HttpServletRequest request, HttpServletResponse response, Object handler) throws Exception {
        System.out.println("执行了 PermissionIntercepter----------preHandle");
        String user=(String) request.getSession().getAttribute("user");
        if(user==null){
```

```
                request.setAttribute("msg", "未经登录,不允许访问,请先登录!");
                request.getRequestDispatcher("/WEB-INF/jsp/login.jsp").forward(request,
response);
                return false;
            }
            return true;
        }
        @Override
        public void afterCompletion(HttpServletRequest request, HttpServletResponse
response, Object handler, Exception arg3)throws Exception {
            System.out.println("执行了PermissionIntercepter----------afterCompletion");
        }
        @Override
        public void postHandle(HttpServletRequest request, HttpServletResponse response,
Object handler, ModelAndView modelAndView)throws Exception {
            System.out.println("执行了PermissionIntercepter----------postHandle");
        }

}
```

（8）注册权限拦截器如下：

```
<!-- 注册拦截器 -->
    <mvc:interceptors>
        <mvc:interceptor>
            <mvc:mapping path="/**"/>
            <mvc:exclude-mapping path="/user/login.do"/>
            <mvc:exclude-mapping path="/user/dologin.do"/>
            <bean class="com.lifeng.intercepter.PermissionIntercepter"/>
        </mvc:interceptor>
    </mvc:interceptors>
```

其中这句话<mvc:exclude-mapping path="/user/login.do"/>是排除登录方法，否则登录也被拦截掉，无法登录。

（9）测试，未登录前，再次访问 http://localhost:8080/springmvc43/user/admin.do，结果如图 11.40 所示。

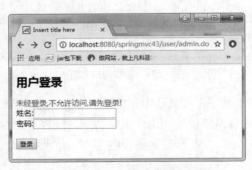

图 11.40 未授权用户无法访问资源

11.8 静态资源访问

如果 web.xml 的配置如下，则核心控制器将拦截一切请求，会导致静态资源无法访问，如访问不了图片和 HTML 静态页面，具体代码如下：

```xml
<servlet>
    <servlet-name>springmvc</servlet-name>
    <servlet-class>org.springframework.web.servlet.DispatcherServlet</servlet-class>
    <init-param>
        <param-name>contextConfigLocation</param-name>
        <param-value>classpath:springmvc.xml</param-value>
    </init-param>
</servlet>
<servlet-mapping>
    <servlet-name>springmvc</servlet-name>
    <url-pattern>/</url-pattern>
</servlet-mapping>
```

要使静态资源能被访问到，必须进行一定的配置，下面介绍几种常用方法。

11.8.1 使用 Tomcat 中名为 default 的 Servlet

在 Tomcat 中有一个专门用于处理静态资源访问的 Servlet：DefaultServlet。其 `<servlet-name/>` 为 default，可以处理各种静态资源访问请求。该 Servlet 注册在 Tomcat 服务器的 web.xml 中，具体是在 Tomcat 安装目录 /conf/web.xml 里面。程序员需要做的就是直接在项目的 web.xml 中注册 `<servlet-mapping/>` 即可。

示例代码如下：

```xml
<servlet-mapping>
    <servlet-name>default</servlet-name>
    <url-pattern>*.jpg</url-pattern>
</servlet-mapping>
<servlet-mapping>
    <servlet-name>default</servlet-name>
    <url-pattern>*.png</url-pattern>
</servlet-mapping>
<servlet-mapping>
    <servlet-name>default</servlet-name>
    <url-pattern>*.js</url-pattern>
</servlet-mapping>
<servlet-mapping>
    <servlet-name>default</servlet-name>
    <url-pattern>*.css</url-pattern>
</servlet-mapping>
```

11.8.2 使用 `<mvc:default-servlet-handler/>`

使用 `<mvc:default-servlet-handler/>`，在 springmvc.xml 中添加 `<mvc:default-servlet-handler/>` 标签即可。

`<mvc:default-servlet-handler/>` 会将静态资源的访问请求添加到 SimpleUrlHandlerMapping 的 urlMap 中，key 就是请求的 URL，value 则为默认 Servlet 请求处理器 DefaultServletHttpRequestHandler 对象。该处理器调用了 Tomcat 的 DefaultServlet 来处理静态资源的访问请求。

要想使用 `<mvc:…/>` 标签，需要引入 MVC 约束。该约束可从 Spring 帮助文档中搜索关键字 spring-mvc.xsd 获取，代码如下：

```xml
<?xml version="1.0" encoding="UTF-8"?>
<beans xmlns="http://www.springframework.org/schema/beans"
    xmlns:xsi="http://www.w3.org/2001/XMLSchema-instance"
    xmlns:mvc="http://www.springframework.org/schema/mvc"
    xmlns:context="http://www.springframework.org/schema/context"
    xmlns:aop="http://www.springframework.org/schema/aop"
    xmlns:tx="http://www.springframework.org/schema/tx"
    xsi:schemaLocation="http://www.springframework.org/schema/beans
        http://www.springframework.org/schema/beans/spring-beans.xsd
        http://www.springframework.org/schema/mvc
        http://www.springframework.org/schema/mvc/spring-mvc.xsd
        http://www.springframework.org/schema/context
        http://www.springframework.org/schema/context/spring-context.xsd
        http://www.springframework.org/schema/aop
        http://www.springframework.org/schema/aop/spring-aop.xsd
        http://www.springframework.org/schema/tx
        http://www.springframework.org/schema/tx/spring-tx.xsd">
    <mvc:default-servlet-handler/>
//省略其他
```

11.8.3 使用\<mvc:resources/\>

在 Spring 3.0.4 版本后，Spring 中定义了专门用于处理静态资源访问请求的处理器 ResourceHttpRequestHandler，并且添加了\<mvc:resources/\>标签，专门用于解决静态资源无法访问的问题。需要在 springmvc.xml 中添加如下形式的配置：

```xml
<mvc:resources location="/images/" mapping="/images/**"/>
<mvc:resources location="/js/" mapping="/js/**"/>
<mvc:resources location="/css/" mapping="/css/**"/>
```

location 表示静态资源所在目录。mapping 表示对该资源的请求路径。要注意，后面是两个星号**。该配置会把对该静态资源的访问请求添加到 SimpleUrlHandlerMapping 的 urlMap 中，key 就是真正与 mapping 的 URL 匹配的 URL，而 value 则为静态资源处理器对象 ResourceHttpRequestHandler。

项目案例：使用静态资源访问。（项目源码参见配套资源：第 11 章/静态资源访问/springmvc44）

关键步骤：

（1）复制项目 springmvc40 为 springmvc44，修改 web.xml 如下：

```xml
<servlet>
    <servlet-name>springmvc</servlet-name>
    <servlet-class>org.springframework.web.servlet.DispatcherServlet</servlet-class>
    <init-param>
        <param-name>contextConfigLocation</param-name>
        <param-value>classpath:springmvc.xml</param-value>
    </init-param>
</servlet>
<servlet-mapping>
    <servlet-name>springmvc</servlet-name>
    <url-pattern>/</url-pattern>  <!-- 拦截一切请求，会导致静态资源无法使用 -->
</servlet-mapping>
```

（2）运行测试，上传图片，结果图片无法显示，如图 11.41 所示。

图 11.41　图片无法显示

这是因为图片是静态资源，被拦截了。

（3）分别采用上述 3 种办法修改 springmvc.xml，代码见上面对应部分。再次测试，效果如图 11.42 所示。

图 11.42　采用 3 种办法修改后又可访问到图片

上机练习

1. 创建一个项目，内有学生类，实现学生信息的数据验证功能。
2. 在上述项目中接着实现学生照片文件的上传与下载。

思考题

1. Spring MVC 处理转发的方式有哪些？
2. Spring MVC 处理重定向的方式有哪些？
3. Spring MVC 如何进行异常处理？
4. 如何实现多种日期格式的转换？
5. 如何实现数据验证？
6. 简述拦截器的执行流程。

第 12 章　Spring MVC 表单标签

本章目标
- 熟悉 Spring 各个标签库的基本语法
- 综合应用 Spring MVC 表单标签

使用 Spring MVC 提供的表单标签可以让 JSP 视图方便地展示 Model 中的数据，特别适合修改现有记录数据的情形，通常进行修改操作时，原有数据要先展示出来，再供用户重新选择或修改。使用表单标签能实现数据绑定，让表单中各个表单域的 name 属性绑定到对象模型中来。JSP 页面中要想使用 SpringMVC 自带标签库，需要导入标签库，在 JSP 文件开头声明如下：

```
<%@ taglib prefix="form" uri="http://www.springframework.org/tags/form" %>
```

12.1　表单标签

Spring 的 form 标签能够自动绑定来自 Model 中的一个属性值到当前 form 对应的实体对象，默认是 command 属性（可以通过 ModelAttribute 重写默认属性），这样就可以在 form 表单体里面方便地使用该对象的属性。form 标签下面又包含 input、password、select/option/options、checkbox/checkboxs、radiobutton、radiobuttons 等子标签。这些标签的基本介绍如表 12.1 所示。

表 12.1　表单标签一览表

标签	作用
form	定义一个表单，是其他子标签的容器，用于生成 HTML 中的<form>标签
input	文本输入框，用于生成 HTML 中的<input type="text">标签
password	密码输入框，用于生成 HTML 中的<input type="password">标签
hidden	隐藏域，用于生成 HTML 中的<input type="hidden">标签
textarea	多行文本域，用于生成 HTML 中的<input type="textarea">标签
checkbox	选择框，用于生成一个 HTML 中的<input type="checkbox">标签
checkboxs	选择框组，用于生成多个 HTML 中的<input type="checkbox">标签
radiobutton	单选框，用于生成一个 HTML 中的<input type="radio">标签
radiobuttons	单选框组，用于生成多个 HTML 中的<input type="radio">标签
select	下拉选择框，用于生成 HTML 中的<select>标签
option	下拉选项，是 select 标签的子标签，用于生成一个 HTML 中的<option>标签
options	下拉选项，是 select 标签的子标签，用于生成多个 HTML 中的<option>标签

12.1.1　form 标签

form 标签主要包括表 12.2 所示的多个属性。

表 12.2　form 标签关键属性

属性	作用
acceptCharset	定义字符编码，如 utf-8
commandName	指定模型对象（Model）的属性名称，如不指定，默认为 command
cssClass	指定需要应用于本表单的 CSS 类
cssStyle	指定需要应用于本表单的 CSS 样式
htmlEscape	设置是否需要进行 HTML 转义，值为 true 或 false
modelAttribute	设置 form backing object 的模型属性名称，如不设置，默认为 command，作用同 commandName

commandName 与 modelAttribute 作用类似，选择其一即可，用于设置该表单绑定的对象。该对象源于控制器中的方法，假设控制器中的方法如下：

```
@RequestMapping("/detail/{id}")
    public ModelAndView detail(@PathVariable int id){
        User myuser=userService.getUserById(id);
        ModelAndView mv=new ModelAndView();
        mv.addObject("user", myuser);
        System.out.println(userService.getUserById(id).getUsername());
        mv.setViewName("detail");
        return mv;
}
```

该代码表示接收到 id 后，调用业务层，获取到该 id 号的 User 类的用户对象 myuser，然后将 myuser 对象封装添加到 ModelAndView 中来，键为"user"，值为 myuser 对象，这样这个模型就封装好了属性名称为 user 的键值对（值为 myuser）。然后进入 JSP 页面，form 表单利用 commandName 与 modelAttribute 指定该模型的"user"属性名，就可绑定该模型属性的值（即 User 类的 myuser 对象），并把该模型属性的值 myuser 对象中的各个属性与 form 表单的各个子标签一一绑定。Form 表单中应用 commandName 绑定模型属性名称示例代码如下：

```
<form:form action="update" method="post" commandName="user">
省略子标签
</form:form>
```

这样就把后台控制器方法中创建的对象传递过来在前台展示了。

12.1.2 input 标签

input 标签，用于展示控制器方法中封装的基本类型的 Model 数据。语法如下：

```
<form:input path="domain 的属性名">
```

其中 path 用于指定要绑定的属性，如上面案例中，表单中指定了要绑定的对象为 User 类的 myuser，但要在表单中具体显示该对象的用户名 username 还需要如下操作：

```
<form:form action="update" method="post" commandName="user">
<form:input path="username"/>
</form:form>
```

这样在表单绑定 myuser 对象的基础上，进一步把该对象的 username 属性绑定到文本输入框中来。

12.1.3 password 标签

password 标签用于填写密码，有掩码，其语法如下：

```
<form:password path="domain 的属性名"/>
```

其 path 属性含义同上。上例中，若还要展示用户的密码，则代码如下所示：

```
<form:form action="update" method="post" commandName="user">
<form:input path="username"/></br>
<form:password path="password"/></br>
</form:form>
```

12.1.4 checkbox 标签

该标签的属性基本与 input 相同，但另有一个 label 属性用于指定显示选框的值，如下所示：

```
<form:checkbox path="domain 的属性名" label="复选框的值">
```

如果该属性名存在，则默认已经选择。后台关键代码如下：

```
myInteresting=new ArrayList<String>();
myInteresting.add("美食");
myInteresting.add("音乐");
```
前台关键代码如下:
```
<form:checkbox path="myInteresting" value="舞蹈" label="舞蹈"/>
<form:checkbox path="myInteresting" value="旅游" label="旅游"/>
<form:checkbox path="myInteresting" value="唱歌" label="唱歌"/>
<form:checkbox path="myInteresting" value="音乐" label="音乐"/>
<form:checkbox path="myInteresting" value="运动" label="运动"/>
<form:checkbox path="myInteresting" value="美食" label="美食"/>
```
显示效果如图 12.1 所示。

图 12.1 多选框效果图

12.1.5 checkboxes 标签

checkboxes 的基本语法如下：

`<form:checkboxes items="${Model 中 Collections 的属性}" path="domain 的属性名"/>`

checkboxes 有几个重要属性，如表 12.3 所示。

表 12.3 checkboxes 的重要属性

属性	作用
delimiter	指定多个多选框之间的分隔符
items	指定用于生成多个多选框的 Collection、Map 或者 Array
itemLabel	为每个多选框提供 label，为 items 属性中定义的 Collection、Map 或者 Array 的对象属性
itemValue	为每个多选框提供 value，为 items 属性中定义的 Collection、Map 或者 Array 的对象属性

假定后台控制器方法中设置了一个集合 allIntersting，用于描述兴趣爱好，其中包含了所有的兴趣爱好：旅游、音乐、唱歌、跳舞、运动、美食。但某一个人的兴趣可能只有其中若干项，所以再定义一个集合 myIntesting，只含部分兴趣爱好。form 表单需要把所有兴趣爱好以多选项的形式列出来，对某一个人有的兴趣爱好则打上钩。后台代码如下：

```
@RequestMapping("/detail/{id}")
    public ModelAndView detail(@PathVariable int id){
        List<String> allInteresting=new ArrayList<String>();
        allInteresting.add("舞蹈");
        allInteresting.add("旅游");
        allInteresting.add("唱歌");
        allInteresting.add("音乐");
        allInteresting.add("运动");
        allInteresting.add("美食");
        UserService userService=new UserService();
        User user=userService.getUserById(id);
        List<String> myInteresting=new ArrayList<String>();
        myInteresting.add("旅游");
        myInteresting.add("音乐");
```

```
            user.setMyInteresting(myInteresting);
            ModelAndView mv=new ModelAndView();
            mv.addObject("allInteresting", allInteresting);
            mv.addObject("user", user);
            mv.setViewName("detail");
            return mv;
```

前台关键代码如下：

```
<form:checkboxes path="myInteresting" items="${allInteresting}"/>
```

最终结果如图 12.2 所示。

图 12.2　checkboxes 多选框效果图

对于 Map 类型的 Collection 来说，key 值为标签的 value 值，value 值为标签的 label 值，但是需要注意 HashMap 是乱序的。

12.1.6　radiobutton 与 radiobuttons 标签

radiobutton 的语法如下：

```
<form:radiobutton path="domain 属性名" label="显示的值" value="值"/>
```

通常有多个值，然后若后台传来的 domain 属性名的值跟其中一个 value 的值相同，则为选中。

radiobuttons 的语法如下：

```
<form:radiobuttons path="domain 属性值" items="${传入的 Collection}"/>
```

其规则同 checkboxes。如果后台传来的 domain 属性名的值跟 items 集合中的其中一个 value 的值相同，则为选中。后台代码如下：

```
@RequestMapping("/detail/{id}")
    public ModelAndView detail(@PathVariable int id){
        List<String> cities=new ArrayList<String>();
        cities.add("北京");
        cities.add("上海");
        cities.add("广州");
        cities.add("深圳");
        String city="深圳";
        UserService userService=new UserService();
        User user=userService.getUserById(id);
        user.setCity(city);
        user.setGender(true);
        ModelAndView mv=new ModelAndView();
        mv.addObject("cities", cities);
        mv.addObject("user", user);
        mv.setViewName("detail");
        return mv;
```

前台代码如下：

```
<tr>
    <td>性别:</td>
    <td>
        <form:radiobutton path="gender" value="true" label="男"/>
        <form:radiobutton path="gender" value="false" label="女"/>
    </td>
```

```
        </tr>
        <tr>
            <td>所在城市 1</td>
            <td>
                <form:radiobuttons path="city" items="${cities}"/><br/>
            </td>
        </tr>
```
最终结果如图 12.3 所示。

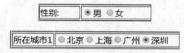

图 12.3 单选框效果图

12.1.7 select 与 option/options 标签

select 与 option/options 标签的语法格式如下：
```
<form:select path="domain 的属性名">
    <form:option value="值">显示的值</form:option>
    <form:options items="${Collections 变量}"/>
</form:select>
```
其中 items 的使用同前。

前台代码同前，后台代码如下：
```
<tr>
    <td>所在城市 2</td>
    <td>
        <form:select path="city">
            <option>请选择城市</option>
            <form:option value="北京">北京</form:option>
            <form:option value="上海">上海</form:option>
            <form:option value="广州">广州</form:option>
            <form:option value="深圳">深圳</form:option>
        </form:select>
    </td>
</tr>
<tr>
    <td>所在城市 3</td>
    <td>
        <form:select path="city">
            <option>请选择城市</option>
            <form:options items="${cities}"/>
        </form:select>
    </td>
</tr>
```
效果如图 12.4 所示。

图 12.4 select 下拉框

12.2 表单标签使用综合案例

项目案例：主页为用户信息列表，单击其中的一项的修改链接，进入该项目的修改页面，原有数据尽数展示。首先展示用户姓名，密码。（项目源码参见本书配套资源：第 12 章/springmvc71）

实现步骤：

（1）将项目 springmvc11 复制为 springmvc71，删除用不到的类。创建实体类 User 如下：

```java
public class User {
    int id;
    String username;
    String password;
    boolean gender;
    String married;
    int age;
    List<String> myInteresting;
    String city;
    String graduated;
    String description;
    //省略 getter,setter 方法
    public User(){
    }
    public User(int id, String username, String password, boolean gender, String married,
int age,List<String> myInteresting, String city, String graduated,String description) {
        super();
        this.id = id;
        this.username = username;
        this.password = password;
        this.gender = gender;
        this.married = married;
        this.age = age;
        this.myInteresting = myInteresting;
        this.city = city;
        this.graduated = graduated;
        this.description = description;
    }
}
```

（2）创建 UserService，模拟从数据库获取 3 条数据，代码如下：

```java
public class UserService {
    List<User> users=new ArrayList<User>();
    //从数据库查找所有 User 用户对象
    public List<User> getUsers() {
        int id=1;
        String username="张无忌";
        String password="123456";
        boolean gender=true;
        String married="是";
        int age=18;
        List<String> myInteresting=new ArrayList<String>();
        myInteresting.add("音乐");
        myInteresting.add("旅游");
        String city="深圳";
```

```java
            String graduated="硕士";
            String description=username+"的个人简介";
            User user1=new User(id, username, password, gender, married,age,
                    myInteresting, city, graduated,description);
            id=2;
            username="李寻欢";
            password="111111";
            gender=false;
            married="否";
            age=19;
            myInteresting=new ArrayList<String>();
            myInteresting.add("运动");
            myInteresting.add("唱歌");
            city="北京";
            graduated="本科";
            description=username+"的个人简介";
            User user2=new User(id, username, password, gender, married,age,
                    myInteresting, city, graduated,description);

            id=3;
            username="黄飞鸿";
            password="888888";
            gender=false;
            married="是";
            age=22;
            myInteresting=new ArrayList<String>();
            myInteresting.add("美食");
            myInteresting.add("音乐");
            city="广州";
            graduated="博士";
            description=username+"的个人简介";
            User user3=new User(id, username, password, gender, married,age,
                    myInteresting, city, graduated,description);
            users.add(user1);
            users.add(user2);
            users.add(user3);
            return users;
    }
    //根据id查找User对象
    public User getUserById(int id) {
        users=getUsers();
        return users.get(id-1);
    }
}
```

（3）创建控制器 UserController，代码如下：

```java
@Controller
@RequestMapping("/user")
public class UserController {
    @RequestMapping("/list")
```

```java
        public ModelAndView list(){
            UserService userService=new UserService();
            List<User> users=userService.getUsers();
            ModelAndView mv=new ModelAndView();
            mv.addObject("users", users);
            mv.setViewName("list");
            return mv;
        }
        @RequestMapping("/detail/{id}")
        public ModelAndView detail(@PathVariable int id){
            List<String> allInteresting=new ArrayList<String>();
            allInteresting.add("舞蹈");
            allInteresting.add("旅游");
            allInteresting.add("唱歌");
            allInteresting.add("音乐");
            allInteresting.add("运动");
            allInteresting.add("美食");
            List<String> cities=new ArrayList<String>();
            cities.add("北京");
            cities.add("上海");
            cities.add("广州");
            cities.add("深圳");
            List<String> graduates=new ArrayList<String>();
            graduates.add("博士");
            graduates.add("硕士");
            graduates.add("本科");
            graduates.add("大专");
            UserService userService=new UserService();
            User user=userService.getUserById(id);
            ModelAndView mv=new ModelAndView();
            mv.addObject("allInteresting", allInteresting);
            mv.addObject("cities", cities);
            mv.addObject("graduates", graduates);
            mv.addObject("user", user);
            mv.setViewName("detail");
            return mv;
        }
    }
```

（4）前台 list.jsp 代码如下，注意要导入 jstl-1.2.jar 包：

```jsp
<%@ page language="java" contentType="text/html; charset=utf-8"
    pageEncoding="utf-8"%>
<%@taglib uri="http://java.sun.com/jsp/jstl/core" prefix="c" %>
<!DOCTYPE html PUBLIC "-//W3C//DTD HTML 4.01 Transitional//EN" "http://www.w3.org/TR/html4/loose.dtd">
<html>
<head>
<meta http-equiv="Content-Type" content="text/html; charset=utf-8">
<title>Insert title here</title>
</head>
<body>
<h2 style="text-align:center">用户列表</h2>
<table>
```

```
<tr>
<td>编号</td><td>姓名</td><td>年龄</td>
<td>详细信息 </td>
</tr>
<c:forEach items="${users }" var="user">
<tr>
<td>${user.id }</td><td>${user.username }</td><td>${user.age }</td>
<td><a href="detail/${user.id }">详细信息 </a></td>
</tr>
</c:forEach>
</table>
</body>
</html>
```

前台 detail.jsp 页面代码如下：

```
<%@ page language="java" contentType="text/html; charset=utf-8"
    pageEncoding="utf-8"%>
    <%@ taglib prefix="form" uri="http://www.springframework.org/tags/form" %>
<!DOCTYPE html PUBLIC "-//W3C//DTD HTML 4.01 Transitional//EN" "http://www.w3.org/TR/html4/loose.dtd">
<html>
<head>
<meta http-equiv="Content-Type" content="text/html; charset=ISO-8859-1">
<title>Insert title here</title>
</head>
<body>
<h1>修改学生信息</h1>
<form:form action="update" method="post" commandName="user">  <!--modelAttribute="user"  commandName="user" -->
    <table border="1">
        <tr>
            <td>用户名:</td>
            <td><form:input path="username"/></td>
        </tr>
        <tr>
            <td>密码:</td>
            <td><form:password path="password"/></td>
        </tr>
        <tr>
            <td>年龄:</td>
            <td><form:input path="age"/></td>
        </tr>
        <tr>
            <td>性别:</td>
            <td>
                <form:radiobutton path="gender" value="true" label="男 "/>
                <form:radiobutton path="gender" value="false" label="女"/>
            </td>
        </tr>
        <tr>
            <td>婚否:</td>
            <td>
                <form:radiobutton path="married" value="是"/>已婚
```

```html
                <form:radiobutton path="married" value="否"/>未婚
            </td>
    </tr>
     <tr>
        <td>兴趣爱好 1</td>
        <td>
            <form:checkbox path="myInteresting" value="舞蹈"/>舞蹈
            <form:checkbox path="myInteresting" value="旅游"/>旅游
            <form:checkbox path="myInteresting" value="唱歌"/>唱歌
            <form:checkbox path="myInteresting" value="音乐"/>音乐
            <form:checkbox path="myInteresting" value="运动"/>运动
            <form:checkbox path="myInteresting" value="美食"/>美食
        </td>
    </tr>
    <tr>
        <td>兴趣爱好 2</td>
        <td>
            <form:checkbox path="myInteresting" value="舞蹈" label="舞蹈"/>
            <form:checkbox path="myInteresting" value="旅游" label="旅游"/>
            <form:checkbox path="myInteresting" value="唱歌" label="唱歌"/>
            <form:checkbox path="myInteresting" value="音乐" label="音乐"/>
            <form:checkbox path="myInteresting" value="运动" label="运动"/>
            <form:checkbox path="myInteresting" value="美食" label="美食"/>
        </td>
    </tr>
    <tr>
        <td>兴趣爱好 3</td>
        <td>
            <form:checkboxes path="myInteresting" items="${allInteresting}" />
        </td>
    </tr>
    <tr>
        <td>所在城市 1</td>
        <td>
            <form:radiobuttons path="city" items="${cities}"/><br/>
        </td>
    </tr>
    <tr>
        <td>所在城市 2</td>
        <td>
            <form:select path="city">
                <option>请选择城市</option>
                    <form:option value="北京">北京</form:option>
                <form:option value="上海">上海</form:option>
                <form:option value="广州">广州</form:option>
                <form:option value="深圳">深圳</form:option>
            </form:select>
        </td>
    </tr>
    <tr>
```

```
                <td>所在城市 3</td>
                <td>
                    <form:select path="city">
                        <option>请选择城市</option>
                        <form:options items="${cities}"/>
                    </form:select>
                </td>
            </tr>
            <tr>
                <td>学历 1</td>
                <td>
                    <form:radiobuttons path="graduated" items="${graduates}"/><br/>
                </td>
            </tr>
            <tr>
                <td>学历 2</td>
                <td>
                    <form:select path="graduated">
                        <option>请选择学历</option>
                        <form:options items="${graduates}"/>
                    </form:select>
                </td>
            </tr>
             <tr>
                <td>个人描述</td>
                <td><form:input path="description"/></td>
            </tr>
            <tr>
                <td colspan="2"><input type="submit" value="提交"/></td>
            </tr>
        </table>
    </form:form>
</body>
</html>
```

（5）访问 http://localhost:8080/springmvc71/user/list，如图 12.5 所示，单击第一项的详细信息，进入 http://localhost:8080/springmvc71/user/detail/1，最终效果如图 12.6 所示。

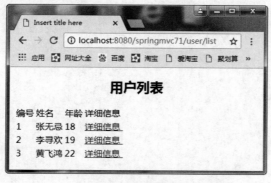

图 12.5　用户列表

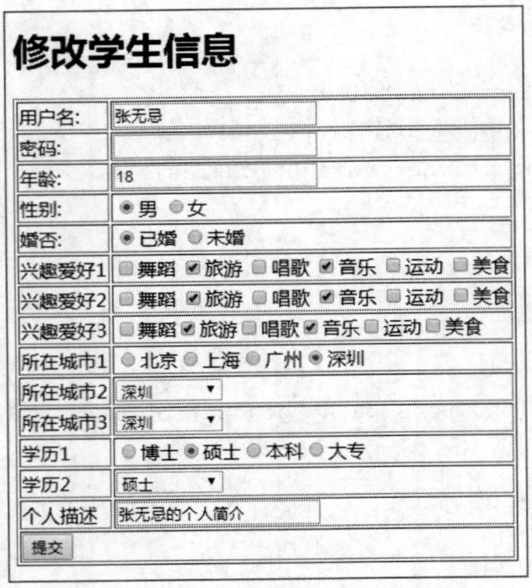

图 12.6 最终效果

上机练习

综合使用本章的各个标签创建一个学生信息修改页面。

思考题

1. form 标签的基本属性有哪些？
2. checkbox 与 checkboxes 的区别是什么？
3. radiobutton 与 radiobuttons 的区别是什么？

第 13 章　SSM 三大框架整合

本章目标

- ◇ 理解 SSM 整合的原理
- ◇ 掌握 SSM 整合环境的搭建流程
- ◇ 掌握 SSM 整合的具体流程

Spring 整合 MyBatis 可以实现让数据库连接、事务管理、实例化对象的创建与依赖关系等都统一由 Spring 负责，以及数据库的增删改查操作由 Spring-MyBatis 整合包提供的 SqlSessionTemplate 对象来操作，或者利用整合包扫描接口，依据 Mapper 映射文件直接创建代理实现类，无须程序员手工实现接口，大大简化了开发流程。Spring MVC 本来就是 Spring 框架的一部分，这两者无须再做整合，所以 SSM 整合的关键就是 Spring 对 MyBatis 的整合，三大框架整合完成后，将以 Spring 为核心，调用有关资源，高效运作。

13.1　Spring 整合 MyBatis

13.1.1　Spring 整合 MyBatis 开发环境

Spring 整合 MyBatis 的开发环境需要 Spring 的 JAR 包和 MyBatis 的 JAR 包，还需要 Spring 与 MyBatis 整合的中间件 mybatis-spring-1.3.1.jar，此外还需要数据库驱动 JAR 包 mysql-connector-java-5.1.37.jar。

13.1.2　DAO 接口实现类开发整合

项目案例：Spring+MyBatis 实现 student 数据库的增删改查。（项目源码参见本书配套资源：第 13 章/Spring 整合 MyBatis/springmybatis1）

实现步骤：

（1）新建项目，导入上述 JAR 包，如图 13.1 所示。

图 13.1　整合 JAR 包

（2）创建实体类 Student 如下：
```
public class Student {
```

```java
        private String sid;
        private String sname;
        private String sex;
        private int age;
        public Student(){
        }
        public Student(String sname,String sex,int age){
            this.sname=sname;
            this.sex=sex;
            this.age=age;
        }
        public Student(String sid,String sname,String sex,int age){
            this.sid=sid;
            this.sname=sname;
            this.sex=sex;
            this.age=age;
        }
        public void show(){
            System.out.println("学生编号:"+sid+" 学生姓名:"+sname+" 学生性别:"+sex+" 学生年龄:"+age);
        }
        //省略setter,getter方法
}
```

（3）创建DAO接口IStudentDao如下：

```java
public interface IStudentDao {
    public void add(Student stu);
    public void delete(int id);
    public void update(Student stu);
    public List<Student> findAllStudents();
    public Student findStudentById(int id);
}
```

（4）创建DAO接口的实现类StudentDaoImpl如下：

```java
public class StudentDaoImpl implements IStudentDao{
    private SqlSessionTemplate sqlSessionTemplate;
    public SqlSessionTemplate getSqlSessionTemplate() {
        return sqlSessionTemplate;
    }
    public void setSqlSessionTemplate(SqlSessionTemplate sqlSessionTemplate) {
        this.sqlSessionTemplate = sqlSessionTemplate;
    }
    @Override
    public void add(Student stu) {
        sqlSessionTemplate.insert("com.lifeng.dao.IStudentDao.add",stu);
    }
    @Override
    public void delete(int id) {
        sqlSessionTemplate.insert("com.lifeng.dao.IStudentDao.delete",id);
    }
    @Override
    public void update(Student stu) {
        sqlSessionTemplate.insert("com.lifeng.dao.IStudentDao.update",stu);
    }
    @Override
    public List<Student> findAllStudents() {
```

```
            List<Student> list=sqlSessionTemplate.selectList("com.lifeng.dao.IStudentDao.
findAllStudents");
            return list;
        }
        @Override
        public Student findStudentById(int id) {
            Student stu=sqlSessionTemplate.selectOne("com.lifeng.dao.IStudentDao.find
StudentById",id);
            return stu;
        }
    }
```

(5)创建业务层接口 IStudentService 如下:
```
public interface IStudentService {
    public void add(Student stu);
    public void delete(int id);
    public void update(Student stu);
    public List<Student> findAllStudents();
    public Student findStudentById(int id);
}
```

(6)创建业务层接口的实现类 StudentServiceImpl 如下:
```
public class StudentServiceImpl implements IStudentService{
    private IStudentDao studentDao;
    public void setStudentDao(IStudentDao studentDao) {
        this.studentDao = studentDao;
    }
    public StudentServiceImpl(){
    }
    @Override
    public void add(Student stu) {
        studentDao.add(stu);
    }
    @Override
    public void delete(int id) {
        studentDao.delete(id);
    }
    @Override
    public void update(Student stu) {
        studentDao.update(stu);
    }
    @Override
    public List<Student> findAllStudents() {
        return studentDao.findAllStudents();
    }
    @Override
    public Student findStudentById(int id) {
        return studentDao.findStudentById(id);
    }
}
```

(7)创建 SQL 映射文件 StudentMapper.xml 如下:
```xml
<?xml version="1.0" encoding="utf-8" ?>
<!DOCTYPE mapper PUBLIC "-//mybatis.org//DTD Mapper 3.0//EN"
"http://mybatis.org/dtd/mybatis-3-mapper.dtd">
<mapper namespace="com.lifeng.dao.IStudentDao">
    <resultMap id="studentResultMap" type="com.lifeng.entity.Student">
```

```xml
            <id property="sid" column="id" />
            <result property="sname" column="studentname" />
            <result property="sex" column="gender" />
            <result property="age" column="age" />
    </resultMap>
    <select id="findAllStudents" resultType="com.lifeng.entity.Student" resultMap="studentResultMap">
        SELECT
            *
        FROM STUDENT
    </select>
    <select id="findStudentById" parameterType="int" resultMap="studentResultMap">
        SELECT
            id,
            studentname,
            gender,
            age
        FROM STUDENT where id=#{id}
    </select>
    <insert id="add" parameterType="Student" >
        INSERT INTO student(id, studentname, gender, age)
        VALUES
          (#{sid}, #{sname}, #{sex}, #{age})
    </insert>
    <delete id="delete" parameterType="int">
       delete from student where id=#{id}
    </delete>
     <update id="update" parameterType="Student">
          UPDATE Student SET studentname=#{sname},gender=#{sex},age=#{age}
          WHERE id= #{sid}
     </update>
</mapper>
```

（8）创建 MyBatis 配置文件 mybatis-config.xml 如下：

```xml
<?xml version="1.0" encoding="UTF-8"?>
<!DOCTYPE configuration PUBLIC "-//mybatis.org//DTD Config 3.0//EN"
"http://mybatis.org/dtd/mybatis-3-config.dtd">
<configuration>
    <typeAliases>
        <typeAlias alias="Student" type="com.lifeng.entity.Student"/>
    </typeAliases>
    <mappers>
        <mapper resource="com/lifeng/dao/StudentMapper.xml"/>
    </mappers>
</configuration>
```

只保留别名，注册映射文件功能，其他数据库连接的功能交给 Spring。

（9）创建 Spring 配置文件 applicationContext.xml 如下：

```xml
<?xml version="1.0" encoding="UTF-8"?>
<beans xmlns="http://www.springframework.org/schema/beans"
   xmlns:xsi="http://www.w3.org/2001/XMLSchema-instance"
   xmlns:aop="http://www.springframework.org/schema/aop"
   xmlns:tx="http://www.springframework.org/schema/tx"
   xmlns:context="http://www.springframework.org/schema/context"
   xsi:schemaLocation="
      http://www.springframework.org/schema/beans
```

```xml
                http://www.springframework.org/schema/beans/spring-beans.xsd
                http://www.springframework.org/schema/context
                http://www.springframework.org/schema/context/spring-context.xsd
                http://www.springframework.org/schema/tx
                http://www.springframework.org/schema/tx/spring-tx.xsd
                http://www.springframework.org/schema/aop
                http://www.springframework.org/schema/aop/spring-aop.xsd">
    <!-- 配置数据源 -->
    <bean id="dataSource" class="org.springframework.jdbc.datasource.DriverManagerDataSource">
        <property name="driverClassName">
            <value>com.mysql.jdbc.Driver</value>
        </property>
        <property name="url">
            <value>jdbc:mysql://localhost:3306/studentdb?characterEncoding=utf8</value>
        </property>
        <property name="username">
            <value>root</value>
        </property>
        <property name="password">
            <value>root</value>
        </property>
    </bean>
    <!-- 定义事务管理器 -->
    <bean id="txManager"
        class="org.springframework.jdbc.datasource.DataSourceTransactionManager">
        <property name="dataSource" ref="dataSource" />
    </bean>
    <!-- 编写通知 -->
    <tx:advice id="txAdvice" transaction-manager="txManager">
        <tx:attributes>
            <tx:method name="*" propagation="REQUIRED"
            isolation="DEFAULT"   read-only="false" />
        </tx:attributes>
    </tx:advice>
    <!-- 编写 AOP,让 Spring 自动切入事务到目标切面 -->
    <aop:config>
        <!-- 定义切入点 -->
        <aop:pointcut id="txPointcut"
            expression="execution(* com.lifeng.service.*.*(..))" />
        <!-- 将事务通知与切入点组合 -->
        <aop:advisor advice-ref="txAdvice" pointcut-ref="txPointcut" />
    </aop:config>
    <!-- 配置 sqlSessionFactory-->
    <bean id="sqlSessionFactory" class="org.mybatis.spring.SqlSessionFactoryBean">
        <property name="dataSource" ref="dataSource" />
        <property name="configLocation" value="classpath:mybatis-config.xml">
</property>
    </bean>
    <!-- 配置 SqlSessionTemplate -->
    <bean id="sqlSessionTemplate" class="org.mybatis.spring.SqlSessionTemplate">
        <constructor-arg name="sqlSessionFactory" ref="sqlSessionFactory" />
    </bean>
    <!-- 配置 DAO 层,注入 SqlSessionTemplate 属性值 -->
```

```xml
    <bean id="studentDao" class="com.lifeng.dao.StudentDaoImpl">
        <property name="sqlSessionTemplate" ref="sqlSessionTemplate"/>
    </bean>
    <!-- 配置SERVICE层,注入studentDao属性值 -->
    <bean id="studentService" class="com.lifeng.service.StudentServiceImpl">
        <property name="studentDao" ref="studentDao"/>
    </bean>
</beans>
```

（10）测试类，代码如下：

```java
public class TestStudent1 {
    public static void main(String[] args) {
        ApplicationContext context=new ClassPathXmlApplicationContext("applicationContext.xml");
        IStudentService stuService=(IStudentService) context.getBean("studentService");
        System.out.println("----------查找全部学生---------");
        List<Student> list=stuService.findAllStudents();
        for(Student stu:list){
            stu.show();
        }
        System.out.println("\n----------查找一个学生---------");
        Student student=stuService.findStudentById(1);
        student.show();
        System.out.println("\n----------添加一个学生---------");
        student=new Student();
        student.setSid("5");
        student.setSname("赵云");
        student.setSex("男");
        student.setAge(22);
        stuService.add(student);
        list=stuService.findAllStudents();
        for(Student stu:list){
            stu.show();
        }
        System.out.println("\n----------修改一个学生---------");
        student=stuService.findStudentById(5);
        student.setSname("赵子龙");
        stuService.update(student);
        list=stuService.findAllStudents();
        for(Student stu:list){
            stu.show();
        }
        System.out.println("\n----------删除一个学生---------");
        stuService.delete(5);
        list=stuService.findAllStudents();
        for(Student stu:list){
            stu.show();
        }
    }
}
```

（11）测试结果如下：

```
----------查找全部学生---------
DEBUG [main] - ==>  Preparing: SELECT * FROM STUDENT
```

```
DEBUG [main] - ==> Parameters:
DEBUG [main] - <==      Total: 4
学生编号:1 学生姓名:张飞 学生性别:男 学生年龄:18
学生编号:2 学生姓名:李白 学生性别:男 学生年龄:20
学生编号:3 学生姓名:张无忌 学生性别:男 学生年龄:19
学生编号:4 学生姓名:赵敏 学生性别:女 学生年龄:17
----------查找一个学生----------
DEBUG [main] - ==>  Preparing: SELECT id, studentname, gender, age FROM STUDENT where id=?
DEBUG [main] - ==> Parameters: 1(Integer)
DEBUG [main] - <==      Total: 1
学生编号:1 学生姓名:张飞 学生性别:男 学生年龄:18
----------添加一个学生----------
DEBUG [main] - ==>  Preparing: INSERT INTO student(id, studentname, gender, age) VALUES (?, ?, ?, ?)
DEBUG [main] - ==> Parameters: 5(String), 赵云(String), 男(String), 22(Integer)
DEBUG [main] - <==    Updates: 1
DEBUG [main] - ==>  Preparing: SELECT * FROM STUDENT
DEBUG [main] - ==> Parameters:
DEBUG [main] - <==      Total: 5
学生编号:1 学生姓名:张飞 学生性别:男 学生年龄:18
学生编号:2 学生姓名:李白 学生性别:男 学生年龄:20
学生编号:3 学生姓名:张无忌 学生性别:男 学生年龄:19
学生编号:4 学生姓名:赵敏 学生性别:女 学生年龄:17
学生编号:5 学生姓名:赵云 学生性别:男 学生年龄:22
----------修改一个学生----------
DEBUG [main] - ==>  Preparing: SELECT id, studentname, gender, age FROM STUDENT where id=?
DEBUG [main] - ==> Parameters: 5(Integer)
DEBUG [main] - <==      Total: 1
DEBUG [main] - ==>  Preparing: UPDATE Student SET studentname=?,gender=?,age=? WHERE id= ?
DEBUG [main] - ==> Parameters: 赵子龙(String), 男(String), 22(Integer), 5(String)
DEBUG [main] - <==    Updates: 1
DEBUG [main] - ==>  Preparing: SELECT * FROM STUDENT
DEBUG [main] - ==> Parameters:
DEBUG [main] - <==      Total: 5
学生编号:1 学生姓名:张飞 学生性别:男 学生年龄:18
学生编号:2 学生姓名:李白 学生性别:男 学生年龄:20
学生编号:3 学生姓名:张无忌 学生性别:男 学生年龄:19
学生编号:4 学生姓名:赵敏 学生性别:女 学生年龄:17
学生编号:5 学生姓名:赵子龙 学生性别:男 学生年龄:22
----------删除一个学生----------
DEBUG [main] - ==>  Preparing: delete from student where id=?
DEBUG [main] - ==> Parameters: 5(Integer)
DEBUG [main] - <==    Updates: 1
DEBUG [main] - ==>  Preparing: SELECT * FROM STUDENT
DEBUG [main] - ==> Parameters:
DEBUG [main] - <==      Total: 4
学生编号:1 学生姓名:张飞 学生性别:男 学生年龄:18
学生编号:2 学生姓名:李白 学生性别:男 学生年龄:20
学生编号:3 学生姓名:张无忌 学生性别:男 学生年龄:19
```

学生编号:4 学生姓名:赵敏 学生性别:女 学生年龄:17

SqlSessionTemplate 的常用方法,如表 13.1 所示。

表 13.1 SqlSessionTemplate 的常用方法

方法	描述
List\<T\> selectList(String statement,Object parameter)	返回结果对象集合,第一个参数 statement 为命名空间+映射 id
T selectOne(String statement,Object paramete)	返回一个结果对象
int insert(String statement,Object paramete)	插入记录,返回插入的行数
int update(String statement,Object paramete)	更新记录,返回更新的行数
int delete(String statement,Object paramete)	删除记录,返回删除的行数
T getMapper(class\<T\> type)	按接口的类型返回接口的实例方法

13.1.3 DAO 接口无实现类开发整合

MyBatis 可以通过代理实现接口,因此不要 DAO 的实现类一样可以做出来。如果没有 DAO 的实现类,业务层 Service 类中的 DAO 属性该如何注入?通过下面这个案例可以得到答案。

项目案例:改造上述项目,去掉 DAO 的实现类,但能实现同样的功能。(项目源码参见本书配套资源:第 13 章/Spring 整合 MyBatis/springmybatis2)

实现步骤:

(1)复制项目 springmybatis1 为 springmybatis2,删除 StudentDaoImpl 类。

(2)修改 Spring 配置文件,删除 SqlSessionTemplate 的配置,修改 DAO 层的配置如下:

```xml
<!-- 配置DAO层,Mapper 代理实现类-->
<bean id="studentDao" class="org.mybatis.spring.mapper.MapperFactoryBean">
    <property name="mapperInterface" value="com.lifeng.dao.IStudentDao"/>
    <property name="sqlSessionFactory" ref="sqlSessionFactory"/>
</bean>
```

这里其实使用了 MapperFactoryBean 类来代理实现接口,假定项目还有其他 DAO 层,一样这样做,依照这个模板再添加一条 Bean 就是了。但如果一个项目有太多的 DAO 类,则需要添加太多的 Bean,可以进行简化,配置批量扫描包,将包下的所有接口自动创建代理实现类,只需下面一条配置就可以,无须一一创建 DAO 层的 Bean:

```xml
<bean class="org.mybatis.spring.mapper.MapperScannerConfigurer">
        <property name="sqlSessionFactory" ref="sqlSessionFactory"/>
        <property name="basePackage" value="com.lifeng.dao"/>
</bean>
```

MapperScannerConfigurer 将扫描 basePackage 属性指定的包下的所有接口类,如果这些接口都有对应的映射文件,则会将它们动态地定义一个 Bean,这样就无须一个个定义 Bean 了。

但问题在于 Service 层,之前有手工创建 DAO 层的 Bean,Service 层再一个个手工注入这些 DAO 层的 Bean,类似下面的配置:

```xml
<!-- 配置DAO层,注入SqlSessionTemplate 属性值 -->
<bean id="studentDao" class="com.lifeng.dao.StudentDaoImpl">
    <property name="sqlSessionTemplate" ref="sqlSessionTemplate"/>
</bean>
<!-- 配置SERVICE层,注入studentDao 属性值 -->
<bean id="studentService" class="com.lifeng.service.StudentServiceImpl">
    <property name="studentDao" ref="studentDao"/>
```

```
</bean>
```

之前这种配置，Service 层注入 studentDao 属性值，关系很清晰，每一个业务 Bean 都能装配具体名称的 DAO 层的 Bean。但现在 studentDao 这个 Bean 配置已经没有了（Bean 由代理动态的自动生成），这样运行的话，程序会报错：

```
Caused by: org.springframework.beans.factory.NoSuchBeanDefinitionException: No bean named 'studentDao' available
    at org.springframework.beans.factory.support.DefaultListableBeanFactory.getBeanDefinition(DefaultListableBeanFactory.java:687)
    at org.springframework.beans.factory.support.AbstractBeanFactory.getMergedLocalBeanDefinition(AbstractBeanFactory.java:1207)
    at org.springframework.beans.factory.support.AbstractBeanFactory.doGetBean(AbstractBeanFactory.java:284)
    at org.springframework.beans.factory.support.AbstractBeanFactory.getBean(AbstractBeanFactory.java:197)
    at org.springframework.beans.factory.support.BeanDefinitionValueResolver.resolveReference(BeanDefinitionValueResolver.java:351)
    ... 15 more
```

提示业务层找不到 studentDao 这个 Bean，解决方案就是让业务层按类型自动装配。业务层的配置代码修改如下：

```xml
<!-- 配置 SERVICE 层,注入 studentDao 属性值 -->
<bean id="studentService" class="com.lifeng.service.StudentServiceImpl" autowire="byType"/>
```

当然，在 Service 层用注解@Autowired 也可以，但要在 Spring 配置文件中开启扫描，具体如下：

```xml
<context:component-scan base-package="com.lifeng.service"/>
```

本案例最终 Spring 的配置文件修改后的完整代码如下：

```xml
<?xml version="1.0" encoding="UTF-8"?>
<beans xmlns="http://www.springframework.org/schema/beans"
    xmlns:xsi="http://www.w3.org/2001/XMLSchema-instance"
    xmlns:aop="http://www.springframework.org/schema/aop"
    xmlns:tx="http://www.springframework.org/schema/tx"
    xmlns:context="http://www.springframework.org/schema/context"
    xsi:schemaLocation="
        http://www.springframework.org/schema/beans
        http://www.springframework.org/schema/beans/spring-beans.xsd
        http://www.springframework.org/schema/context
        http://www.springframework.org/schema/context/spring-context.xsd
        http://www.springframework.org/schema/tx
        http://www.springframework.org/schema/tx/spring-tx.xsd
        http://www.springframework.org/schema/aop
        http://www.springframework.org/schema/aop/spring-aop.xsd">
    <!-- <context:component-scan base-package="com.lifeng.service"/> -->
    <!-- 配置数据源 -->
    <bean id="dataSource" class="org.springframework.jdbc.datasource.DriverManagerDataSource">
        <property name="driverClassName">
            <value>com.mysql.jdbc.Driver</value>
        </property>
        <property name="url">
            <value>jdbc:mysql://localhost:3306/studentdb?characterEncoding=utf8</value>
        </property>
        <property name="username">
            <value>root</value>
```

```xml
        </property>
        <property name="password">
            <value>root</value>
        </property>
    </bean>
    <!-- 定义事务管理器 -->
    <bean id="txManager"
        class="org.springframework.jdbc.datasource.DataSourceTransactionManager">
        <property name="dataSource" ref="dataSource" />
    </bean>
    <!-- 编写通知 -->
    <tx:advice id="txAdvice" transaction-manager="txManager">
        <tx:attributes>
            <tx:method name="*" propagation="REQUIRED"
            isolation="DEFAULT"  read-only="false" />
        </tx:attributes>
    </tx:advice>
    <!-- 编写 AOP,让 Spring 自动切入事务到目标切面 -->
    <aop:config>
        <!-- 定义切入点 -->
        <aop:pointcut id="txPointcut"
            expression="execution(* com.lifeng.service.*.*(..))" />
        <!-- 将事务通知与切入点组合 -->
        <aop:advisor advice-ref="txAdvice" pointcut-ref="txPointcut" />
    </aop:config>
    <!-- 配置sqlSessionFactory-->
    <bean id="sqlSessionFactory" class="org.mybatis.spring.SqlSessionFactoryBean">
        <property name="dataSource" ref="dataSource" />
        <property name="configLocation" value="classpath:mybatis-config.xml"></property>
    </bean>
    <!-- 配置 DAO 层,Mapper 代理实现类-->
    <!--<bean id="studentDao" class="org.mybatis.spring.mapper.MapperFactoryBean">
        <property name="mapperInterface" value="com.lifeng.dao.IStudentDao"/>
        <property name="sqlSessionFactory" ref="sqlSessionFactory"/>
    </bean> -->
    <bean class="org.mybatis.spring.mapper.MapperScannerConfigurer">
        <property name="sqlSessionFactory" ref="sqlSessionFactory"/>
        <property name="basePackage" value="com.lifeng.dao"/>
    </bean>
    <!-- 配置 SERVICE 层,注入 studentDao 属性值 -->
    <bean id="studentService" class="com.lifeng.service.StudentServiceImpl" autowire="byType"/>
</beans>
```

(3)其他不变,测试结果同前。本案例的最大特点就是没有 DAO 的实现类,但同样可以实现程序的所有功能。

13.2 SSM 整合案例

项目案例:SSM 整合 Web 项目,实现学生信息的基本管理。(项目源码参见本书配套资源:第

13 章/SSM 整合/ssm1）

实现步骤：

（1）新建项目 ssm1，在 13.1 节整合的所需 JAR 包的基础上，再把 Spring MVC 的有关 JAR 包导入项目，总的 JAR 包如图 13.2 所示。

图 13.2　SSM 整合所需的 JAR 包

（2）配置 web.xml 文件。指定 Spring 配置文件的位置如下：

```xml
<!-- 注册 Spring 配置文件的位置 -->
<context-param>
    <param-name>contextConfigLocation</param-name>
    <param-value>classpath:applicationContex.xml</param-value>
</context-param>
```

注册 ServletContext 监听器如下：

```xml
<!-- 注册 ServletContext 监听器,创建 Spring 容器 -->
<listener>
    <listener-class>
        org.springframework.web.context.ContextLoaderListener
    </listener-class>
</listener>
```

（3）注册字符集过滤器，解决请求参数的中文乱码问题，代码如下：

```xml
<!-- 注册字符集过滤器-->
<filter>
        <filter-name>characterEncodingFilter</filter-name>
        <filter-class>org.springframework.web.filter.CharacterEncodingFilter</filter-class>
        <init-param>
            <param-name>encoding</param-name>
```

```xml
            <param-value>UTF-8</param-value>
        </init-param>
        <init-param>
            <param-name>forceEncoding</param-name>
            <param-value>true</param-value>
        </init-param>
    </filter>
    <filter-mapping>
        <filter-name>characterEncodingFilter</filter-name>
        <url-pattern>/*</url-pattern>
    </filter-mapping>
```

(4) 配置 Spring-MVC 核心控制器，代码如下：

```xml
<!-- 配置 Spring 核心控制器 -->
  <servlet>
        <servlet-name>springmvc</servlet-name>
        <servlet-class>org.springframework.web.servlet.DispatcherServlet</servlet-class>
        <init-param>
             <param-name>contextConfigLocation</param-name>
             <param-value>classpath:springmvc.xml</param-value>
        </init-param>
  </servlet>
  <servlet-mapping>
        <servlet-name>springmvc</servlet-name>
        <url-pattern>*.do</url-pattern>
  </servlet-mapping>
```

web.xml 的完整配置如下：

```xml
<?xml version="1.0" encoding="UTF-8"?>
<web-app xmlns:xsi="http://www.w3.org/2001/XMLSchema-instance" xmlns="http://xmlns.jcp.org/xml/ns/javaee" xsi:schemaLocation="http://xmlns.jcp.org/xml/ns/javaee http://xmlns.jcp.org/xml/ns/javaee/web-app_3_1.xsd" id="WebApp_ID" version="3.1">
    <!-- 注册 Spring 配置文件的位置 -->
    <context-param>
        <param-name>contextConfigLocation</param-name>
        <param-value>classpath:applicationContex.xml</param-value>
    </context-param>
    <!-- 注册 ServletContext 监听器,创建 Spring 容器 -->
    <listener>
        <listener-class>
            org.springframework.web.context.ContextLoaderListener
        </listener-class>
    </listener>
    <!-- 注册字符集过滤器-->
    <filter>
        <filter-name>characterEncodingFilter</filter-name>
        <filter-class>org.springframework.web.filter.CharacterEncodingFilter</filter-class>
        <init-param>
            <param-name>encoding</param-name>
            <param-value>UTF-8</param-value>
        </init-param>
        <init-param>
            <param-name>forceEncoding</param-name>
            <param-value>true</param-value>
```

```xml
            </init-param>
        </filter>
        <filter-mapping>
            <filter-name>characterEncodingFilter</filter-name>
            <url-pattern>/*</url-pattern>
        </filter-mapping>
        <!-- 配置 Spring 核心控制器 -->
    <servlet>
        <servlet-name>springmvc</servlet-name>
        <servlet-class>org.springframework.web.servlet.DispatcherServlet</servlet-class>
        <init-param>
            <param-name>contextConfigLocation</param-name>
            <param-value>classpath:springmvc.xml</param-value>
        </init-param>
    </servlet>
    <servlet-mapping>
        <servlet-name>springmvc</servlet-name>
        <url-pattern>*.do</url-pattern>
    </servlet-mapping>
    <welcome-file-list>
      <welcome-file>index.jsp</welcome-file>
    </welcome-file-list>
</web-app>
```

(5) 在 src 下创建 Spring 配置文件 applicationContext.xml。注册数据源 DataSource 如下:

```xml
<!-- 配置数据源 c3p0 数据源 -->
<bean id="dataSource" class="com.mchange.v2.c3p0.ComboPooledDataSource">
    <property name="driverClass">
        <value>com.mysql.jdbc.Driver</value>
    </property>
    <property name="jdbcUrl">
        <value>jdbc:mysql://localhost:3306/studentdb</value>
    </property>
    <property name="user">
        <value>root</value>
    </property>
    <property name="password">
        <value>root</value>
    </property>
</bean>
```

注册事务如下:

```xml
<!-- 定义事务管理器 -->
    <bean id="txManager"
        class="org.springframework.jdbc.datasource.DataSourceTransactionManager">
        <property name="dataSource" ref="dataSource" />
    </bean>
    <!-- 编写通知 -->
    <tx:advice id="txAdvice" transaction-manager="txManager">
        <tx:attributes>
            <tx:method name="*" propagation="REQUIRED"
        isolation="DEFAULT"    read-only="false" />
        </tx:attributes>
    </tx:advice>
```

```xml
<!-- 编写AOP,让Spring自动切入事务到目标切面 -->
<aop:config>
    <!-- 定义切入点 -->
    <aop:pointcut id="txPointcut"
        expression="execution(* com.lifeng.service.*.*(..))" />
    <!-- 将事务通知与切入点组合 -->
    <aop:advisor advice-ref="txAdvice" pointcut-ref="txPointcut" />
</aop:config>
```

注册 MyBatis 配置文件如下：

```xml
<!-- 配置sqlSessionFactory-->
<bean id="sqlSessionFactory" class="org.mybatis.spring.SqlSessionFactoryBean">
    <property name="dataSource" ref="dataSource" />
    <!-- 指定mybatis配置文件的位置 -->
    <property name="configLocation" value="classpath:mybatis-config.xml"></property>
</bean>
<!-- 注册Mapper扫描配置器 -->
<bean class="org.mybatis.spring.mapper.MapperScannerConfigurer">
    <property name="sqlSessionFactory" ref="sqlSessionFactory"/>
    <property name="basePackage" value="com.lifeng.dao"/>
</bean>
```

这样配置无须创建 DAO 层接口的实现类，项目会自动生成 DAO 层接口的代理实现类。

配置注解扫描 service，以注解的形式创建 Bean 如下：

```xml
<!-- 配置扫描service层 -->
<context:component-scan base-package="com.lifeng.service"/>
```

到此，Spring 配置文件 applicationContext.xml 完整代码如下：

```xml
<?xml version="1.0" encoding="UTF-8"?>
<beans xmlns="http://www.springframework.org/schema/beans"
    xmlns:xsi="http://www.w3.org/2001/XMLSchema-instance"
    xmlns:aop="http://www.springframework.org/schema/aop"
    xmlns:tx="http://www.springframework.org/schema/tx"
    xmlns:context="http://www.springframework.org/schema/context"
    xsi:schemaLocation="
        http://www.springframework.org/schema/beans
        http://www.springframework.org/schema/beans/spring-beans.xsd
        http://www.springframework.org/schema/context
        http://www.springframework.org/schema/context/spring-context.xsd
        http://www.springframework.org/schema/tx
        http://www.springframework.org/schema/tx/spring-tx.xsd
        http://www.springframework.org/schema/aop
        http://www.springframework.org/schema/aop/spring-aop.xsd">
    <!-- 配置数据源c3p0数据源 -->
    <bean id="dataSource" class="com.mchange.v2.c3p0.ComboPooledDataSource">
        <property name="driverClass">
            <value>com.mysql.jdbc.Driver</value>
        </property>
        <property name="jdbcUrl">
            <value>jdbc:mysql://localhost:3306/studentdb</value>
        </property>
        <property name="user">
            <value>root</value>
        </property>
```

```xml
                <property name="password">
                    <value>root</value>
                </property>
            </bean>
            <!-- 定义事务管理器 -->
            <bean id="txManager"
                class="org.springframework.jdbc.datasource.DataSourceTransactionManager">
                <property name="dataSource" ref="dataSource" />
            </bean>
            <!-- 编写通知 -->
            <tx:advice id="txAdvice" transaction-manager="txManager">
                <tx:attributes>
                    <tx:method name="*" propagation="REQUIRED"
                isolation="DEFAULT"   read-only="false" />
                </tx:attributes>
            </tx:advice>
            <!-- 编写AOP,让Spring自动切入事务到目标切面 -->
            <aop:config>
                <!-- 定义切入点 -->
                <aop:pointcut id="txPointcut"
                    expression="execution(* com.lifeng.service.*.*(..))" />
                <!-- 将事务通知与切入点组合 -->
                <aop:advisor advice-ref="txAdvice" pointcut-ref="txPointcut" />
            </aop:config>
            <!-- 配置sqlSessionFactory-->
            <bean id="sqlSessionFactory" class="org.mybatis.spring.SqlSessionFactoryBean">
                <property name="dataSource" ref="dataSource" />
                <!-- 指定mybatis配置文件的位置 -->
                <property name="configLocation" value="classpath:mybatis-config.xml">
</property>
            </bean>
            <!-- 注册Mapper扫描配置器 -->
            <bean class="org.mybatis.spring.mapper.MapperScannerConfigurer">
                <property name="sqlSessionFactory" ref="sqlSessionFactory"/>
                <property name="basePackage" value="com.lifeng.dao"/>
            </bean>
            <!-- 配置扫描service层 -->
            <context:component-scan base-package="com.lifeng.service"/>
</beans>
```

（6）在 src 下创建 MyBatis 配置文件 mybatis-config.xml，代码如下：

```xml
<?xml version="1.0" encoding="UTF-8"?>
<!DOCTYPE configuration PUBLIC "-//mybatis.org//DTD Config 3.0//EN"
"http://mybatis.org/dtd/mybatis-3-config.dtd">
<configuration>
    <typeAliases>
        <typeAlias alias="Student" type="com.lifeng.entity.Student"/>
    </typeAliases>
    <mappers>
        <mapper resource="com/lifeng/dao/StudentMapper.xml"/>
    </mappers>
</configuration>
```

（7）在 src 下创建 Spring 配置文件 springmvc.xml，代码如下：

```xml
<?xml version="1.0" encoding="UTF-8"?>
```

```xml
<beans xmlns="http://www.springframework.org/schema/beans"
    xmlns:xsi="http://www.w3.org/2001/XMLSchema-instance"
    xmlns:mvc="http://www.springframework.org/schema/mvc"
    xmlns:context="http://www.springframework.org/schema/context"
    xmlns:aop="http://www.springframework.org/schema/aop"
    xmlns:tx="http://www.springframework.org/schema/tx"
    xsi:schemaLocation="http://www.springframework.org/schema/beans
        http://www.springframework.org/schema/beans/spring-beans.xsd
        http://www.springframework.org/schema/mvc
        http://www.springframework.org/schema/mvc/spring-mvc.xsd
        http://www.springframework.org/schema/context
        http://www.springframework.org/schema/context/spring-context.xsd
        http://www.springframework.org/schema/aop
        http://www.springframework.org/schema/aop/spring-aop.xsd
        http://www.springframework.org/schema/tx
        http://www.springframework.org/schema/tx/spring-tx.xsd">
    <!-- 配置包扫描器,扫描@Controller注解的类 -->
    <context:component-scan base-package="com.lifeng.controller"/>
    <!-- 加载注解驱动 -->
    <mvc:annotation-driven />
    <!-- 配置视图解析器 -->
    <bean class="org.springframework.web.servlet.view.InternalResourceViewResolver">
        <!--逻辑视图前缀-->
        <property name="prefix" value="/WEB-INF/jsp/"></property>
        <!--逻辑视图后缀,匹配模式:前缀+逻辑视图+后缀,形成完整路径名-->
        <property name="suffix" value=".jsp"></property>
    </bean>
</beans>
```

这样 SSM 框架基本就配置好了，即可进行应用开发。

（8）创建数据库表，具体代码如下：

```
create table student (
    id double ,
    studentname varchar (60),
    gender varchar (6),
    age double
);
insert into student (id, studentname, gender, age) values('1','张飞','男','18');
insert into student (id, studentname, gender, age) values('2','李白','男','20');
insert into student (id, studentname, gender, age) values('3','张无忌','男','19');
insert into student (id, studentname, gender, age) values('4','赵敏','女','17');
```

（9）新建包 com.lifeng.entity 创建实体类 Student，具体代码如下：

```java
public class Student {
    private String sid;
    private String sname;
    private String sex;
    private int age;
    public Student(){
    }
    public Student(String sname,String sex,int age){
        this.sname=sname;
        this.sex=sex;
        this.age=age;
```

```java
    }
    public Student(String sid,String sname,String sex,int age){
        this.sid=sid;
        this.sname=sname;
        this.sex=sex;
        this.age=age;
    }
    public void show(){
        System.out.println("学生编号:"+sid+" 学生姓名:"+sname+" 学生性别:"+sex+" 学生年龄:"+age);
    }
    //省略getter,setter方法
}
```

（10）创建包 com.lifeng.dao，创建接口 IstudentDao，代码如下：

```java
public interface IStudentDao {
    public void add(Student stu);
    public void delete(int id);
    public void update(Student stu);
    public List<Student> findAllStudents();
    public Student findStudentById(int id);
}
```

（11）在包 com.lifeng.dao 下创建接口 IStudentDao 对应的映射文件 StudentMapper.xml，代码如下：

```xml
<?xml version="1.0" encoding="utf-8" ?>
<!DOCTYPE mapper PUBLIC "-//mybatis.org//DTD Mapper 3.0//EN"
"http://mybatis.org/dtd/mybatis-3-mapper.dtd">
<mapper namespace="com.lifeng.dao.IStudentDao">
    <resultMap id="studentResultMap" type="com.lifeng.entity.Student">
        <id property="sid" column="id" />
        <result property="sname" column="studentname" />
        <result property="sex" column="gender" />
        <result property="age" column="age" />
    </resultMap>
    <select id="findAllStudents" resultType="com.lifeng.entity.Student" resultMap="studentResultMap">
        SELECT
            *
        FROM STUDENT
    </select>
    <select id="findStudentById" parameterType="int" resultMap="studentResultMap">
        SELECT
            id,
            studentname,
            gender,
            age
        FROM STUDENT where id=#{id}
    </select>
    <!-- 主键自增长 -->
    <insert id="add" parameterType="Student" keyProperty="sid" useGeneratedKeys="true">
        INSERT INTO student(studentname, gender, age)
        VALUES
            (#{sname}, #{sex}, #{age})
    </insert>
    <delete id="delete" parameterType="int">
```

```xml
        delete from student where id=#{id}
    </delete>
    <update id="update" parameterType="Student">
        UPDATE Student SET studentname=#{sname},gender=#{sex},age=#{age}
        WHERE id= #{sid}
    </update>
</mapper>
```

（12）创建包 com.lifeng.service，创建接口 IstudentService，代码如下：

```java
public interface IStudentService {
    public void add(Student stu);
    public void delete(int id);
    public void update(Student stu);
    public List<Student> findAllStudents();
    public Student findStudentById(int id);
    public void testTransaction();
}
```

（13）创建接口 IStudentService 的实现类 StudentServiceImpl，添加注解，代码如下：

```java
@Service
public class StudentServiceImpl implements IStudentService{
    @Autowired
    private IStudentDao studentDao;
    public void setStudentDao(IStudentDao studentDao) {
        this.studentDao = studentDao;
    }
    public StudentServiceImpl(){
    }
    @Override
    public void add(Student stu) {
        studentDao.add(stu);
    }
    @Override
    public void delete(int id) {
        studentDao.delete(id);
    }
    @Override
    public void update(Student stu) {
        studentDao.update(stu);
    }
    @Override
    public List<Student> findAllStudents() {
        return studentDao.findAllStudents();
    }
    @Override
    public Student findStudentById(int id) {
        return studentDao.findStudentById(id);
    }
    public void testTransaction(){
        delete(4);
        Integer.parseInt("a");
        delete(3);
    }
}
```

注解@Service 的作用是创建这个类的一个实例化 Bean。

注解@Autowired 的作用是给属性 studentDao 自动装配，默认按类型装配。

（14）创建包 com.lifeng.controller，创建控制器类，代码如下：

```
@Controller
public class StudentController{
    @Autowired
    IStudentService studentService;
    @RequestMapping("/add.do")
    public ModelAndView add(Student student) {
        ModelAndView mv=new ModelAndView();
        studentService.add(student);
        mv.setViewName("redirect:findAll.do");
        return mv;
    }
    @RequestMapping("/delete.do")
    public ModelAndView delete(int id) {
        ModelAndView mv=new ModelAndView();
        studentService.delete(id);
        mv.setViewName("redirect:findAll.do");
        return mv;
    }
    @RequestMapping("/update.do")
    public ModelAndView update(Student student) {
        ModelAndView mv=new ModelAndView();
        studentService.update(student);
        mv.setViewName("redirect:findAll.do");
        return mv;
    }
    @RequestMapping("/findAll.do")
    public ModelAndView findAll() {
        ModelAndView mv=new ModelAndView();
        List<Student> list=studentService.findAllStudents();
        mv.addObject("list",list);
        mv.setViewName("list");
        return mv;
    }
    @RequestMapping("/findSingle.do")
    public ModelAndView findSingle(int id) {
        ModelAndView mv=new ModelAndView();
        Student stu=studentService.findStudentById(id);
        mv.addObject("stu",stu);
        mv.setViewName("detail");
        return mv;
    }
    @RequestMapping("/toUpdate.do")
    public ModelAndView toUpdate(int id) {
        ModelAndView mv=new ModelAndView();
        Student stu=studentService.findStudentById(id);
        mv.addObject("stu",stu);
        mv.setViewName("update");
        return mv;
    }
}
```

（15）在目录 WebContent/WEB-INF 下新建文件夹 jsp，创建以下页面。

add.jsp 代码如下：
```
<body>
<br/>
<br/>
<h2 style="text-align:cneter">添加学生</h2>
<form action="add.do" method="post">
姓名:<input type="text" name="sname" /></br>
性别:<input type="text" name="sex" /></br>
年龄:<input type="text" name="age" /></br>
<input type="submit" value="添加"/>
</form>
</body>
```

detail.jsp 代码如下：
```
<body>
<h2 style="text-align:center">学生详细信息</h2>
<table>
<tr><td>编号:</td><td>${stu.sid}</td></tr>
<tr><td>姓名:</td><td>${stu.sname}</td></tr>
<tr><td>性别:</td><td>${stu.sex}</td></tr>
<tr><td>年龄:</td><td>${stu.age}</td></tr>
<tr><td colspan=2><a href="findAll.do">返回</a></td></tr>
</table>
</body>
```

list.jsp 代码如下：
```
<body>
<h2 style="text-align:center">学生列表</h2>
<table style="border:1;cellspacing:0">
<tr>
<td>编号:</td><td>姓名:</td><td>性别:</td><td>年龄:</td>
<td>详细信息 </td><td>修改 </td><td>删除 </td>
</tr>
<c:forEach item="${list}" var="stu">
<tr>
<td>${stu.sid}</td><td>${stu.sname}</td><td>${stu.sex}</td><td>${stu.age}</td>
<td><a href="detail.do?id=${stu.sid}">详细信息 </a></td>
<td><a href="toUpdate.do?id=${stu.sid}">修改 </a></td>
<td><a href="delete.do?id=${stu.sid}" onclick="return confirm('你确认要删除该学生信息吗?')">删除 </a></td>
</tr>
</c:forEach>
</table>
</body>
```

update.jsp 代码如下：
```
<body>
<br/>
<br/>
<h2 style="text-align:cneter">添加学生</h2>
<form action="update.do" method="post">
姓名:<input type="text" name="sname" value="${stu.sname}"/></br>
```

```
性别:<input type="text" name="sex" value="${stu.sname}"/></br>
年龄:<input type="text" name="age" value="${stu.sname}"/></br>
<input type="hidden" name="sid" value="${stu.sid}"/>
<input type="submit" value="修改"/>
</form>
</body>
```

（16）运行测试效果。学生列表如图 13.3 所示。

图 13.3　学生列表

添加学生如图 13.4 所示。

图 13.4　添加学生

修改学生如图 13.5 所示。

图 13.5　修改学生

最终项目结构如图 13.6 所示。

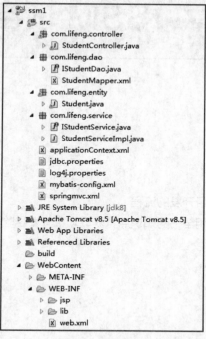

图 13.6　项目结构

上机练习

1. Spring 整合 MyBatis，查找年龄小于等于 19 岁的学生。
2. 模拟银行转账，用事务管理解决张三的钱转出去了，李四的钱却没转进来的情况。
3. 用 SSM 框架对商品信息表进行增删改查。

思考题

1. SSM 整合的原理是什么？
2. SSM 整合的流程是怎样的？
3. Spring 整合 MyBatis 后，哪些功能由 Spring 来实现？哪些功能仍由 MyBatis 来实现？

第 14 章　SSM 项目实战

本章目标
- ◇ 掌握项目开发的流程
- ◇ 综合利用全书的知识点，获取项目开发经验

14.1 项目需求分析

项目名称：砺锋网上服装城。

功能需求：

（1）顾客能够登录、自动登录，可以浏览分类商品及商品详情，可以按动态条件搜索商品、查看商品评论、加入购物车、管理购物车、下订单、模拟支付等，实现顾客购物的完整流程。

（2）顾客进入后台管理订单。

（3）网店管理员管理商品与订单。

本书提供详细步骤实现上述功能（1），这也是本项目最重要的功能。其余功能作为课后作业，留给读者自行完善。（项目源码参见本书配套资源）

技术要求：Spring 4.3.4+Spring MVC+MyBatis 3.4.5；

数据库：MySQL 5.7；

前端框架：jQuery+Bootstrap；

服务器：Tomcat 8.5；

Java 版本：JDK 1.8；

开发工具：Eclipse 4.7.2。

14.2 搭建 SSM 框架

（1）在 Eclipse 中新建 Web 项目 dress，导入 MyBatis 所需的 JAR 包及 Spring4.3.4 所需的 JAR 包。其他 JAR 包还有：连接 MySQL 所需的 JAR 包 mysql-connector-java-5.1.38.jar、JSON 所需的 JAR 包。全部所需 JAR 包如图 14.1 所示（JAR 包参见本书配套资源：第 14 章/dress）。

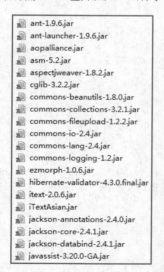

图 14.1 项目所需的全部 JAR 包

这里还导入了 poi、itext 等 JAR 包，以便生成 Excel 或 PDF 文件时用。

（2）新建 mybatis-config.xml 配置文件，只需定义别名和指定映射文件即可，关键代码如下：

```xml
<configuration>
<typeAliases>
    <typeAlias alias="User" type="com.lifeng.entity.User"/>
</typeAliases>
<mappers>
    <mapper resource="com/lifeng/dao/UserMapper.xml"/>
</mappers>
</configuration>
```

以上内容需要根据功能模块的添加而增加。

（3）新建 springmvc.xml 配置文件，关键代码如下：

```xml
<!-- 配置包扫描器,扫描@Controller 注解的类 -->
<context:component-scan base-package="com.lifeng.controller"/>
<!-- 加载注解驱动 -->
<mvc:annotation-driven/>
<!-- 配置静态资源访问,匹配的文件将不受前端控制器拦截 -->
<mvc:default-servlet-handler/>
<!-- 配置视图解释器,代码参考随书资源源文件-->
```

（4）在 src 下新建 applicationContext.xml 配置文件，关键代码如下：

```xml
<!-- 配置数据源 代码见源文件-->
<!-- 定义事务管理器 代码见源文件-->
<!-- 编写通知 代码见源文件-->
<!-- 编写 AOP,让 Spring 自动切入事务到目标切面 代码见源文件-->
<!-- 配置 sqlSessionFactory-->
<bean id="sqlSessionFactory" class="org.mybatis.spring.SqlSessionFactoryBean">
    <property name="dataSource" ref="dataSource" />
    <!-- 指定 mybatis 配置文件的位置 -->
    <property name="configLocation" value="classpath:mybatis-config.xml">
</property>
</bean>
<!-- 注册 Mapper 扫描配置器 -->
<bean class="org.mybatis.spring.mapper.MapperScannerConfigurer">
    <property name="sqlSessionFactory" ref="sqlSessionFactory"/>
    <property name="basePackage" value="com.lifeng.dao"/>
</bean>
<!-- 配置扫描 service 层 -->
<context:component-scan base-package="com.lifeng.service"/>
</beans>
```

（5）配置 web.xml，代码如下：

```xml
<?xml version="1.0" encoding="UTF-8"?>
<web-app xmlns:xsi="http://www.w3.org/2001/XMLSchema-instance" xmlns="http://xmlns.jcp.org/xml/ns/javaee" xsi:schemaLocation="http://xmlns.jcp.org/xml/ns/javaee http://xmlns.jcp.org/xml/ns/javaee/web-app_3_1.xsd" id="WebApp_ID" version="3.1">
    <display-name>dress</display-name>
<!-- 注册 Spring 配置文件的位置 -->
<context-param>
    <param-name>contextConfigLocation</param-name>
    <param-value>classpath:applicationContext.xml</param-value>
</context-param>
<!-- 注册 ServletContext 监听器,创建 Spring 容器 -->
```

```xml
<listener>
    <listener-class>
        org.springframework.web.context.ContextLoaderListener
    </listener-class>
</listener>
<!-- 注册字符集过滤器-->
<filter>
    <filter-name>characterEncodingFilter</filter-name>
    <filter-class>org.springframework.web.filter.CharacterEncodingFilter</filter-class>
    <init-param>
        <param-name>encoding</param-name>
        <param-value>UTF-8</param-value>
    </init-param>
    <init-param>
        <param-name>forceEncoding</param-name>
        <param-value>true</param-value>
    </init-param>
</filter>
<filter-mapping>
    <filter-name>characterEncodingFilter</filter-name>
    <url-pattern>/*</url-pattern>
</filter-mapping>
<!-- 配置 Spring 核心控制器 -->
<servlet>
    <servlet-name>springmvc</servlet-name>
    <servlet-class>org.springframework.web.servlet.DispatcherServlet</servlet-class>
    <init-param>
        <param-name>contextConfigLocation</param-name>
        <param-value>classpath:springmvc.xml</param-value>
    </init-param>
    <load-on-startup>1</load-on-startup>
</servlet>
<servlet-mapping>
    <servlet-name>springmvc</servlet-name>
    <url-pattern>/</url-pattern>
</servlet-mapping>
<welcome-file-list>
  <welcome-file>index.jsp</welcome-file>
</welcome-file-list>
</web-app>
```

（6）新建包 com.lifeng.util，在包下新建一个 Unicode.java 工具类，用于解决传递参数中的中文乱码问题。这个类只是辅助类，所以不再列出，请参考配套资源。

14.3 首页与用户登录模块设计

在浏览器中访问 http://localhost:8080/dress/index，出现首页，首页效果如图 14.2 所示。

图 14.2 首页

将光标移动到左侧主题后,其右侧即会弹出半透明的子菜单,如图 14.3 所示。

图 14.3 分类导航

单击右上角的"登录"按钮后,系统会弹出登录框,如图 14.4 所示。

图 14.4 弹出登录框

用户登录成功后,右上角及右侧登录状态会发生改变,如图 14.5 所示。

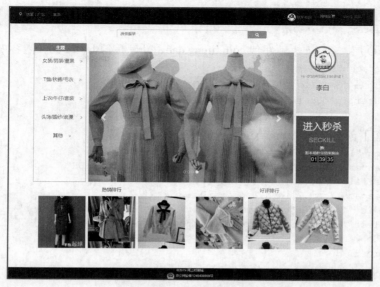

图 14.5　登录成功后的首页

实现步骤：

（1）在 MySQL 中创建数据库 dressdb，创建表 user，如图 14.6 所示。

Alter Table 'user' in 'dressdb'

Field Name	Datatype	Len	Default	PK?	Not Null?	Unsigned?	Auto Incr?	Zerofill?
id	int	11		✓	✓		✓	
username	varchar	20			✓			
password	varchar	20			✓			
gender	varchar	10						
email	varchar	50			✓			
telephone	varchar	20			✓			
introduce	varchar	100						
role	varchar	10	ordinary		✓			
registTime	timestamp		CURRENT_TIMESTAMP		✓			
shippingAddress	varchar	1000						
name	varchar	20			✓			
headimg	varchar	255	img/userheadimg/0		✓			

Alter　Advanced Properties...　Preview...　Insert　Delete　Cancel　Hide Language Options

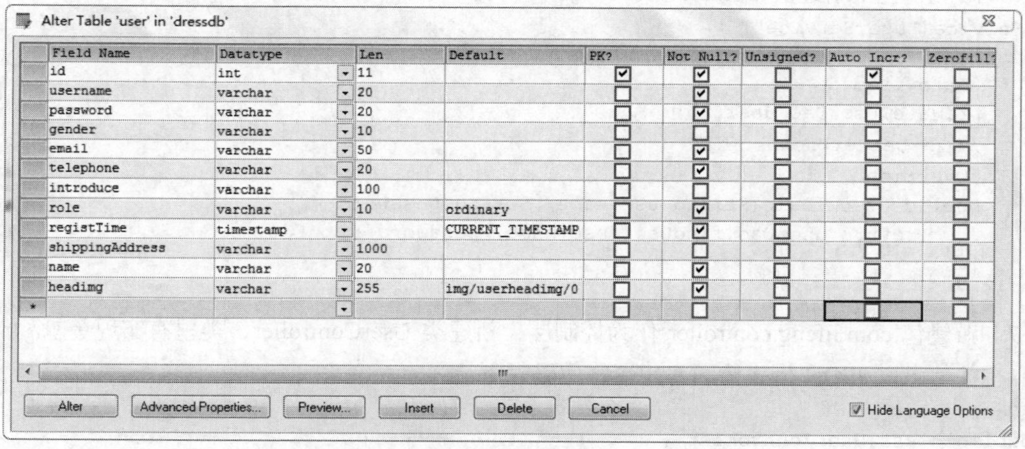

图 14.6　user 结构

（2）新建包 com.lifeng.entity，新建实体类 User，代码如下：

```
public class User implements Serializable{
    private int id;
    private String username;
    private String password;
    private String gender;
    private String email;
    private String telephone;
    private String introduce;
    private String role;
    private Date registTime;
    private String shippingAddress;
    private String name;
```

```
            private String headimg;
        //省略getter,setter方法
```

（3）创建DAO层，在src下新建包com.lifeng.dao，新建接口UserDao，内容如下：
```
public interface UserDao {
    // 根据用户名与密码查找用户
    public User findUser(String username, String password);
}
```

（4）在同一包下创建UserMapper.xml映射文件，代码如下：
```
<?xml version="1.0" encoding="utf-8" ?>
<!DOCTYPE mapper PUBLIC "-//mybatis.org//DTD Mapper 3.0//EN"
"http://mybatis.org/dtd/mybatis-3-mapper.dtd">
<mapper namespace="com.lifeng.dao.UserDao">
    <select id="findUser" parameterType="String" resultType="User">
        select * from user where username=#{0} and password=#{1}
    </select>
</mapper>
```

（5）新建包com.lifeng.service作为业务层，新建接口UserService如下：
```
public interface UserService {
    // 用户登录
    public User login(String username, String password);
}
```

（6）同一个包下新建类UserServiceImpl.java实现UserService接口。
```
@Service("userService")
public class UserServiceImpl implements UserService{
    @Autowired
    private UserDao userDao;
    //省略getter,setter方法
    @Override
    public User login(String username, String password) {
        return userDao.findUser(username, password);
    }
}
```

（7）新建包com.lifeng.controller作为控制层，新建类UserController，类里有实现登录和退出的方法。相关信息以JSON格式传回客户端，代码如下：
```
@Controller
public class UserController {
    @Autowired
    private UserService userService;
    //省略getter,setter方法
    //用于访问主页
    @RequestMapping("/index")
    public ModelAndView index() {
        ModelAndView mv=new ModelAndView();
        mv.setViewName("index");
        return mv;
    }
    //用于登录
    @RequestMapping("/login")
    @ResponseBody
    public Object login(HttpSession session,HttpServletResponse response,String username,String password,String autologin) {
```

```java
            System.out.println("username:"+username+" password:"+password);
            User user = userService.login(username, password);
            Map<String,Object> map = new HashMap<String,Object>();
            try {
                //登录成功,存入cookie,以便下次自动登录
                if(user != null) {
                    session.setAttribute("user", user);
                    if(autologin.equals("true")) {
                        Cookie cookie = new Cookie("user", new Unicode().toCookieUnicode(user.getUsername()) + "==" +new Unicode().toCookieUnicode(user.getPassword()));
                        cookie.setPath("/");
                        cookie.setMaxAge(1000*60*60*24);//24 小时
                        response.addCookie(cookie);
                    }
                    map.put("state","true");
                    map.put("user", user);
                }else {
                    map.put("state","false");
                }
            }catch(Exception e){
                e.printStackTrace();
            }
            return map;
    }
    //用于退出登录
    @RequestMapping("/logout")
    public void logout(HttpSession session,HttpServletResponse response) throws IOException, ServletException {
        session.removeAttribute("user");
        //清除 cookie
        Cookie cookie = new Cookie("user", "");
        cookie.setPath("/");
        cookie.setMaxAge(0);
        response.addCookie(cookie);
        response.sendRedirect("index");
    }
}
```

（8）实现自动登录。

上面代码已实现若用户登录时勾选了自动登录,则服务器会创建有效期为 24 小时的 cookie 存入客户端,下次再访问 index.jsp 页面时可以通过拦截器进行自动登录。首先在 src 下新建一个包 com.lifeng.interceptor,在包下新建类 AutoLoginInterceptor 继承 AbstractInterceptor,代码如下:

```java
public class AutoLoginInterceptor implements HandlerInterceptor{
    //省略其他方法
    @Override
    public boolean preHandle(HttpServletRequest request, HttpServletResponse response, Object arg2) throws Exception {
        Cookie[] cookies = request.getCookies();
        String username = "";
        String password = "";
        //为避免中文乱码问题,使用了工具类 Unicode,在包 com.bookshop.util 下
        Unicode unicode = new Unicode();
        //遍历 cookie
```

```
                if(cookies !=null) {
                    for(int i = 0; i <cookies.length; i++) {
                        if(cookies[i].getName().equals("user")) {
                            username = unicode.toCookieString(cookies[i].getValue().split("==")[0]);
                            password = unicode.toCookieString(cookies[i].getValue().split("==")[1]);
                            break;
                        }
                    }
                    if((username !=null || username !="") && (password !=null || password !="")) {
                        ApplicationContext applicationContext=
                            new ClassPathXmlApplicationContext("applicationContext.xml");
                        UserService userService=applicationContext.getBean("userService", UserService.class);
                        User user=userService.login(username, password);
                        //如果是合法用户,则设置 session
                        if(user != null) {
                            request.getSession().setAttribute("user", user);
                        }
                    }
                }
                return true;
            }
        }
```

（9）修改 spring.mvc 文件，定义和使用拦截器。关键代码如下：

```
<!-- 拦截器,用于自动登录 -->
    <mvc:interceptors>
        <mvc:interceptor>
            <!-- 拦截所有操作 -->
            <mvc:mapping path="/**"/>
            <!-- 把登录操作排除掉 -->
            <mvc:exclude-mapping path="/login"/>
            <bean class="com.lifeng.interceptor.AutoLoginInterceptor"/>
        </mvc:interceptor>
    </mvc:interceptors>
```

首次登录时若勾选了自动登录，则存入了 cookie 到客户端，下次此客户端再次访问 http://localhost:8080/dress/index，则拦截器会进行拦截，获取 cookie 进行验证，验证通过，则创建 session，从而实现自动登录。

【注意】下次访问首页 URL 必须是 index，不能是 index.jsp，否则拦截器拦截不到，无法实现自动登录。

（10）在 WebContent 下建立以下文件夹。
- bootstrap：放置 bootstrap 资源。
- img：放置各个页面要用到的图片。
- css：放置各个页面要用到的 css 文件。
- js：放置各个页面要用到的 JS 文件。
- jspt：放置重复使用的页面头部和脚部，供其他页面调用。

读者仿做此项目时，可直接从随书资源中复制以上所有文件夹。

jspt 文件夹里面包含了 navbar.jsp 文件，作为供其他页面重复调用的头部。navbar.jsp 文件包含了登录模块，但一开始让它隐藏，只有登录后才可使用 jQuery 的下拉特效显示出来。登录使用 Ajax，传递用户名和密码到控制器，传回登录状态、用户对象等，再在 Ajax 回调函数 success 中改变若干页面元素，使之显示用户名、头像等。具体代码见随书资源。

（11）创建 index.jsp，导入前端资源，将 navbar.jsp 包含进来。首页还有图片轮播功能，可使用 Bootstrap 实现。关键代码及完整代码请参见配套资源。接下来就可以测试运行了，在浏览器中输入 http://localhost:8080/dress/index，可以看到首页，单击"登录"按钮，弹出登录框，输入用户名和密码，选择自动登录，登录成功后页面右上角会显示用户名。关闭浏览器，再次输入 url：http://localhost:8080/dress/index，可发现目前处于自动登录状态。

首页还有分类查看商品和搜索商品功能，留到下一个模块再进行介绍。

14.4 商品查询与分页模块设计

本项目共有三处搜索业务。首页有一个搜索框，搜索条件只有一个，就是服装名称，首页还有分类（主题）查询服装，查询条件也只有一个，就是服装的种类 category。此外，分类后的 lisp.jsp 页面有多条件动态查询，查询条件有多个。这里将它们统一为多条件动态查询来处理，不论条件有多少个，搜索完毕都用分页呈现。

（1）在 MySQL 数据库中创建表，如图 14.7 所示。

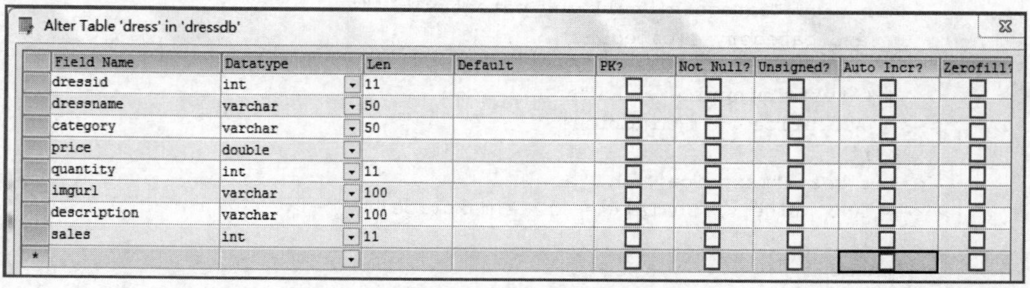

Field Name	Datatype	Len	Default	PK?	Not Null?	Unsigned?	Auto Incr?	Zerofill?
dressid	int	11		☐	☐	☐	☐	☐
dressname	varchar	50		☐	☐	☐	☐	☐
category	varchar	50		☐	☐	☐	☐	☐
price	double			☐	☐	☐	☐	☐
quantity	int	11		☐	☐	☐	☐	☐
imgurl	varchar	100		☐	☐	☐	☐	☐
description	varchar	100		☐	☐	☐	☐	☐
sales	int	11		☐	☐	☐	☐	☐

图 14.7　dress 表结构

在包 com.lifeng.entity 下新建对应的实体类 Dress 代码如下：

```
public class Dress implements Serializable{
private int dressid;
private String dressname;
private double price;
private String category;
private int quantity;
private String imgurl;
private String description;
private int sales;
//省略getter,setter方法
}
```

（2）在包 com.lifeng.dao 下新建 DressMapper.xml 映射文件，代码如下：

```
<?xml version="1.0" encoding="utf-8" ?>
```

```xml
<!DOCTYPE mapper PUBLIC "-//mybatis.org//DTD Mapper 3.0//EN"
"http://mybatis.org/dtd/mybatis-3-mapper.dtd">
<mapper namespace="com.lifeng.dao.DressDao">
    <select id="findDressPage" parameterType="java.util.HashMap" resultType="Dress">
        select * from dress
        <where>
          <if test="dressname!=null and dressname!=''">
            and dressname like '%' #{dressname} '%'
          </if>
          <if test="category!=null and category!=''">
            and category=#{category}
          </if>
          <if test="minprice > 0.0">
            AND price <![CDATA[ >= ]]> #{minprice}
          </if>
          <if test="maxprice > 0.0">
            AND price<![CDATA[ <= ]]>#{maxprice}
          </if>
        </where>
        limit #{startRow},#{pageSize}
    </select>
    <select id="findDressCount" parameterType="java.util.HashMap" resultType="int">
        SELECT count(*)   FROM dress
        <where>
          <if test="dressname!=null and dressname!=''">
            and dressname like '%' #{dressname} '%'
          </if>
          <if test="category!=null and category!=''">
            and category=#{category}
          </if>
          <if test="minprice > 0.0">
            AND price >= #{minprice}
          </if>
          <if test="maxprice > 0.0">
            AND price <![CDATA[ <= ]]> #{maxprice}
          </if>
        </where>
    </select>
</mapper>
```

然后修改 mybatis-config.xml 文件，添加内容如下：

```xml
<configuration>
    <typeAliases>
        <typeAlias alias="User" type="com.lifeng.entity.User"/>
        <typeAlias alias="Dress" type="com.lifeng.entity.Dress"/>
    </typeAliases>
    <mappers>
        <mapper resource="com/lifeng/dao/UserMapper.xml"/>
        <mapper resource="com/lifeng/dao/DressMapper.xml"/>
    </mappers>
</configuration>
```

（3）在包 com.lifeng.entity 下新建实体类 DressCondition，用于封装查询条件，代码如下：

```java
public class DressCondition implements Serializable{
    private String dressname;
    private double minprice;
```

```java
        private double maxprice;
        private String category;
        //省略getter,setter方法
        public DressCondition() {}
        public DressCondition(String dressname, double minprice, double maxprice, String category) {
            super();
            //省略赋值语句
        }
    }
```

（4）在包 com.lifeng.entity 下新建实体类 PageBean，用于封装分页信息，代码如下：

```java
public class PageBean {
    private int currentPage;
    private int pageSize;
    private int count;
    private int totalPage;
    private List<Dress> dresses;
    //封装查询条件
    private DressCondition dressCondition;
    //省略getter,setter方法
}
```

（5）在 com.lifeng.dao 下新建接口 DressDao，代码如下：

```java
public interface DressDao {
    // 根据查询条件查询时装并分页
    public List<Dress> findDressPage(Map<String, Object> map);
    //根据查询条件查询记录条数
    public int findDressCount(Map<String, Object> map);
}
```

（6）在 com.lifeng.service 下新建接口 DressService，代码如下：

```java
public interface DressService {
    // 获取所有服装并分页
    public PageBean findDressPage(int currentPage, int pageSize, String dressname, String category,double minprice, double maxprice);
}
```

（7）在 com.lifeng.service 下新建接口 DressService 的实现类 DressServiceImpl，代码如下：

```java
@Service("dressService")
public class DressServiceImpl implements DressService{
    @Autowired
    private DressDao dressDao;
    //省略getter,setter方法
    @Override
    public PageBean findDressPage(int currentPage, int pageSize, String dressname, String category,double minprice, double maxprice) {
        PageBean pb = new PageBean();
        pb.setCurrentPage(currentPage);
        pb.setPageSize(pageSize);
        try {
            Map<String,Object> map=new HashMap<String,Object>();
            map.put("dressname", dressname);
            map.put("category", category);
            map.put("minprice", minprice);
            map.put("maxprice", maxprice);
```

```
                    int count = dressDao.findDressCount(map);//得到总记录数
                    pb.setCount(count);
                    int totalPage = (int)Math.ceil(count*1.0/pageSize); //求出总页数
                    pb.setTotalPage(totalPage);
                    map.put("startRow", (currentPage-1)*pageSize);
                    map.put("pageSize", pageSize);
                    List<Dress> dresses= dressDao.findDressPage(map);
                    pb.setDresses(dresses);
                    DressCondition  dressCondition=new  DressCondition(dressname,minprice,
maxprice, category);
                    System.out.println("dressCondition 里面的 category:"+dressCondition.
getCategory());
                    pb.setDressCondition(dressCondition);
            } catch (Exception e) {
                e.printStackTrace();
            }
            return pb;
        }
    }
```

（8）在包 com.lifeng.controller 下新建类 DressController，代码如下：

```
@Controller
public class DressController {
    @Autowired
    private DressService dressService;
    //省略 getter,setter 方法

    @RequestMapping("/page")
    public ModelAndView page(int currentPage, String dressname, String category, double
minprice, double maxprice) {
        ModelAndView mv = new ModelAndView();
        // 分页查询,并返回 PageBean 对象
        try {
            int pageSize = 4;// 暂定每页显示 4 条
            PageBean pb = dressService.findDressPage(currentPage, pageSize, dressname,
category, minprice, maxprice);
            mv.addObject("pb", pb);
            mv.setViewName("list");
        } catch (Exception e) {
            e.printStackTrace();
        }
        return mv;
    }
}
```

（9）给 index.jsp 页面的搜索框按钮添加 JS 点击事件。直接在 index.jsp 页面底部添加以下代码：

```
<script type="text/javascript">
    /* 搜索按钮提交事件 */
    $("#sosoBtn").click(function(){
        var t = $("#soso").val();
        window.location.href="page?currentPage=1&minprice=0.0&maxprice=0.0&
dressname="+t+"&category=";
    });
</script>
```

（10）index.jsp 页面的主题导航栏的第一项超链接，其实际类别为"女装"，如图 14.8 所示。代码如下：

```
<a href="page?category=女装&currentPage=1&minprice=0.0&maxprice=0.0"><font>女装/男装/童装</font></a>
```

图 14.8　分类导航

（11）在 WebContent 下新建 list.jsp 页面，以显示搜索结果，代码参见随书资源。

（12）进行测试，在首页单击左侧分类导航，"女装/男装/童装"类（这里实际查询的类别为"女装"），显示图 14.9 所示效果，可以分页。

图 14.9　分类分页显示服装

（13）测试搜索框，多条件查询同样可以。结果如图 14.10 所示。

图 14.10　查询结果

14.5 商品详情模块设计

单击图 14.9 中的某个服装的图像，将进入商品详情页面。可以查看商品的详细信息，同时查看对该商品的评论，如图 14.11 所示。

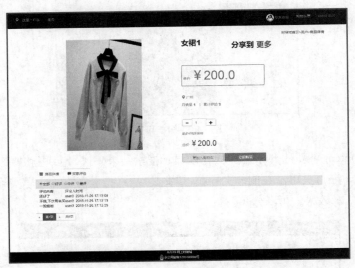

图 14.11 商品详情页面

实现步骤：

（1）在 MySQL 中创建表 comment，用于记录评论。表结构如图 14.12 所示。

Alter Table 'comment' in 'dressdb'

Field Name	Datatype	Len	Default	PK?	Not Null?	Unsigned?	Auto Incr?	Zerofill?
id	int	20		☑	☑	☐	☑	☐
orderid	int	20		☐	☐	☐	☐	☐
comments	varchar	255		☐	☐	☐	☐	☐
dressid	int	20		☐	☑	☐	☐	☐
score	int	1		☐	☑	☐	☐	☐
userid	int	20		☐	☑	☐	☐	☐
time	timestamp		CURRENT_TIMESTAMP	☐	☑	☐	☐	☐
imgurl	varchar	255		☐	☐	☐	☐	☐

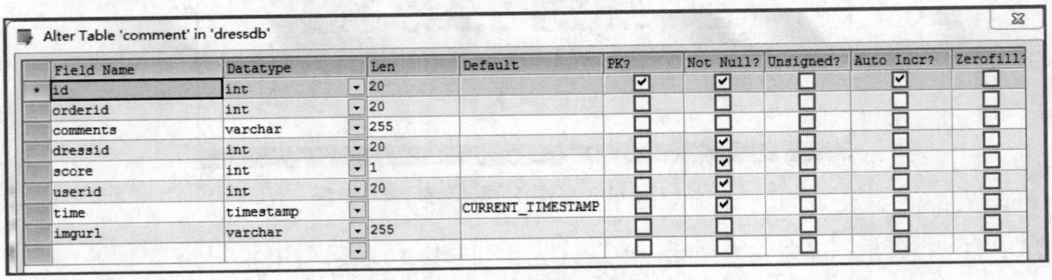

图 14.12 表 comment 的结构

在包 com.lifeng.entity 下新建对应的实体类 Comment，评论跟用户是多对一的关系，即一个用户可发表多个评论，可针对一件服装发表多个评论，代码如下：

```
public class Comment {
private int id;
private int orderid;
private String dressid;
private String comments;
private int score;
private User user;
private Date time;
private String imgurl;
//省略 getter,setter 方法
}
```

（2）在 com.lifeng.dao 包下创建 CommentMapper.xml 映射文件，代码如下：

```xml
<?xml version="1.0" encoding="utf-8" ?>
<!DOCTYPE mapper PUBLIC "-//mybatis.org//DTD Mapper 3.0//EN"
"http://mybatis.org/dtd/mybatis-3-mapper.dtd">
<mapper namespace="com.lifeng.dao.CommentDao">
    <resultMap id="commentResultMap" type="Comment">
        <id property="id" column="id" />
        <result property="comments" column="comments" />
        <result property="score" column="score" />
        <result property="time" column="time" />
        <result property="imgurl" column="imgurl" />
        <result property="orderid" column="orderid" />
        <result property="dressid" column="dressid" />
        <!-- 关联属性 -->
        <association property="user" javaType="User">
            <id property="id" column="id" />
            <result property="username" column="username" />
        </association>
    </resultMap>
    <select id="findComment" parameterType="int" resultMap="commentResultMap">
        select c.id,c.orderid,c.comments,c.dressid,c.score,c.userid,
        c.time,c.imgurl,u.id,u.username
        from comment c,user u
        where c.userid=u.id
        and c.dressid=#{dressid}
    </select>
</mapper>
```

修改 mybatis-config.xml 如下：

```xml
<configuration>
    <typeAliases>
        <typeAlias alias="User" type="com.lifeng.entity.User"/>
        <typeAlias alias="Dress" type="com.lifeng.entity.Dress"/>
        <typeAlias alias="Comment" type="com.lifeng.entity.Comment"/>
    </typeAliases>
    <mappers>
        <mapper resource="com/lifeng/dao/UserMapper.xml"/>
        <mapper resource="com/lifeng/dao/DressMapper.xml"/>
        <mapper resource="com/lifeng/dao/CommentMapper.xml"/>
    </mappers>
</configuration>
```

（3）在包 com.lifeng.dao 下创建 CommentDao 接口，代码如下：

```java
public interface CommentDao {
    // 以服装id寻找评论
    public List<Comment> findComment(int dressid);
}
```

（4）在 DressDao 接口中添加方法，代码如下：

```java
// 以id获取服装
public Dress findDressById(int id);
```

在 DressMapper.xml 中添加方法，代码如下：

```xml
<select id="findDressById" parameterType="int" resultType="Dress">
    select * from dress where dressid=#{0}
</select>
```

（5）在业务层的 DressService 接口中添加方法，代码如下：
```
// 以 id 获取服装
public Dress findDressById(int id);
```
（6）在业务层的 DressServiceImpl 类中添加方法，实现上述接口，代码如下：
```
    @Override
    public Dress findDressById(int id) {
        return dressDao.findDressById(id);
    }
```
（7）在包 com.lifeng.service 下新建 CommentService 接口，代码如下：
```
public interface CommentService {
// 以服装 id 获取评论
    public List<Comment> findComment(int dressid);
}
```
（8）在包 com.lifeng.service 下新建 CommentServiceImpl 类，实现上述接口，代码如下：
```
    @Service("commentService")
    public class CommentServiceImpl implements CommentService{
        @Autowired
        private CommentDao commentDao;
        //省略 getter/setter
        // 以服装 id 查找评论
        @Override
        public List<Comment> findComment(int dressid) {
            return commentDao.findComment(dressid);
        }
    }
```
（9）在包 com.lifeng.controller 下的 DressController 中添加方法 showDress()，代码如下：
```
@Autowired
private CommentService commentService;
//省略 getter/setter
    @RequestMapping("/showDress")
    public ModelAndView showDress(int id) {
        ModelAndView mv=new ModelAndView();
        Dress dress=dressService.findDressById(id);
        List<Comment> evlist=commentService.findComment(id);
        mv.addObject("dress",dress);
        mv.addObject("evlist",evlist);
        mv.setViewName("dress");
        return mv;
    }
```
（10）在 WebContent 下创建 dress.jsp 页面，用于展示商品详情。其中，图片放大、添加与减少购买数量、添加购物车等功能都可利用 JS 实现，跟它配套的 JS 文件是 dress.js，请参见配套资源。

（11）运行测试，可正常浏览商品详情页面，并显示相关评论。

14.6 购物车模块设计

商品详情页面 book.jsp 中已有添加购物车的按钮，以 Ajax 请求的方式，实现将商品信息保存进数据库 cart 表，JS 代码会先检查用户是否登录，只有登录了才能添加到购物车。商品查询分页列表 list.jsp 也有添加到购物车的超链接，功能类似。还可以在 book.jsp 页面右上角的购物车链接中查看当前用户的购物车信息。下面介绍其完整过程。

（1）在 MySQL 中创建数据库表 cart，如图 14.13 所示。

Alter Table 'cart' in 'dressdb'

Field Name	Datatype	Len	Default	PK?	Not Null?	Unsigned?	Auto Incr?
id	int	20		☑	☑	☐	☑
dressid	int	20		☐	☑	☐	☐
userid	int	20		☐	☑	☐	☐
booknum	int	11		☐	☐	☐	☐

图 14.13　表 cart 的结构

包 com.lifeng.entity 中新建实体类 Cart.java，代码如下：

```
public class Cart implements Serializable{
private int id;
    private User user;
    private Dress dress;
    private int booknum;  //省略 getter,setter 方法
}
```

（2）在 DAO 层 com.lifeng.dao 包中新建接口 CartDao，代码如下：

```
public interface CartDao {
    // 获取某个用户的所有购物车
    public List<Cart> findCartByUserId(int userid);
    // 获取用户的某个商品的购物车
    public Cart findCart(int dressid, int userid);
    // 以购物车 id 寻找商品
    public Cart findCartByCartId(int id);
    // 删除单个购物车
    public void removeCart(int id);
    // 添加到购物车
    public void addCart(Cart cart);
    // 修改购物车服装数量
    public void updateCart(Cart cart);
}
```

（3）在 DAO 层中新建 CartMapper.xml 映射文件，代码如下：

```
<?xml version="1.0" encoding="utf-8" ?>
<!DOCTYPE mapper PUBLIC "-//mybatis.org//DTD Mapper 3.0//EN"
"http://mybatis.org/dtd/mybatis-3-mapper.dtd">
<mapper namespace="com.lifeng.dao.CartDao">
    <resultMap id="cartResultMap" type="Cart">
        <id property="id" column="id" />
        <result property="booknum" column="booknum" />
        <!-- 关联属性 -->
        <association property="dress" javaType="Dress">
            <id property="dressid" column="dressid" />
            <result property="dressname" column="dressname" />
            <result property="price" column="price" />
            <result property="imgurl" column="imgurl" />
            <result property="quantity" column="quantity" />
```

```xml
            <result property="sales" column="sales" />
        </association>
        <association property="user" javaType="User">
            <id property="id" column="id" />
            <result property="username" column="username" />
        </association>
    </resultMap>
    <select id="findCartByUserId" parameterType="int" resultMap="cartResultMap">
        select c.id,c.dressid,c.userid,c.booknum,u.id,u.username,d.dressid,d.dressname,d.imgurl,d.price,d.quantity,d.sales from cart c,user u,dress d
        where c.dressid=d.dressid and c.userid=u.id
        and c.userid=#{userid}
    </select>
    <select id="findCart" parameterType="int" resultMap="cartResultMap">
        select c.id,c.dressid,c.userid,c.booknum,u.id,u.username,d.dressid,d.dressname,d.imgurl,d.price,d.quantity,d.sales from cart c,user u,dress d
        where c.dressid=d.dressid and c.userid=u.id
        and c.userid=#{0} and c.dressid=#{1}
    </select>
    <select id="findCartByCartId" parameterType="int" resultMap="cartResultMap">
        select c.id,c.dressid,c.userid,c.booknum,u.id,u.username,d.dressid,d.dressname,d.imgurl,d.price,d.quantity,d.sales from cart c,user u,dress d
        where c.dressid=d.dressid and c.userid=u.id
        and c.id=#{id}
    </select>
    <delete id="removeCart" parameterType="int">
        delete from cart where id=#{id}
    </delete>
    <insert id="addCart" parameterType="Cart" >
        INSERT INTO cart(dressid, userid, booknum)
        VALUES
          (#{dress.dressid}, #{user.id}, #{booknum})
    </insert>
    <update id="updateCart" parameterType="Cart">
        UPDATE cart SET booknum=#{booknum} WHERE id= #{id}
    </update>
</mapper>
```

修改 mybatis-config.xml 文件，代码如下：

```xml
<configuration>
    <typeAliases>
        <typeAlias alias="User" type="com.lifeng.entity.User"/>
        <typeAlias alias="Dress" type="com.lifeng.entity.Dress"/>
        <typeAlias alias="Comment" type="com.lifeng.entity.Comment"/>
        <typeAlias alias="Cart" type="com.lifeng.entity.Cart"/>
    </typeAliases>
    <mappers>
        <mapper resource="com/lifeng/dao/UserMapper.xml"/>
        <mapper resource="com/lifeng/dao/DressMapper.xml"/>
        <mapper resource="com/lifeng/dao/CommentMapper.xml"/>
        <mapper resource="com/lifeng/dao/CartMapper.xml"/>
    </mappers>
</configuration>
```

(4) 在业务层中新建接口 CartService，代码如下：
```java
public interface CartService {
// 获取用户的所有购物车
public List<Cart> findCartByUserId(int userid);
// 寻找购物车
public Cart findCart(int dressid, int userid);
// 以购物车 id 查的购物车
public Cart findCartByCartId(int id);
// 删除单个购物车
public void removeCart(int id) ;
// 添加到购物车
public void addCart(Cart cart);
// 修改购物车服装数量
public void updateCart(Cart cart);
}
```

(5) 在业务层中新建 CartServiceImpl.java，实现上述接口，代码如下：
```java
@Service("cartService")
public class CartServiceImpl implements CartService{
    @Autowired
    private CartDao cartDao;
    //省略 getter/setter
    @Override
    public List<Cart> findCartByUserId(int id) {
        return cartDao.findCartByUserId(id);
    }
    @Override
    public Cart findCart(int products_id, int user_id) {
        return cartDao.findCart(products_id, user_id);
    }
    @Override
    public void removeCart(int id) {
        cartDao.removeCart(id);
    }
    @Override
    public void addCart(Cart cart) {
        cartDao.addCart(cart);
    }
    @Override
    public Cart findCartByCartId(int id) {
        return cartDao.findCartByCartId(id);
    }
    @Override
    public void updateCart(Cart cart) {
        cartDao.updateCart(cart);
    }
}
```

(6) 在包 com.lifeng.controller 下新建 CartController，代码如下：
```java
@Controller
public class CartController{
    @Autowired
    private CartService cartService;
    @Autowired
```

```java
        private DressService dressService;
        //省略getter/setter
        //添加购物车
        @RequestMapping("/addCart")
        @ResponseBody
        public Object addCart(HttpSession session,int dressid,int bookSum) {
            Map<String,Object> map=new HashMap<String,Object>();
            User user=(User) session.getAttribute("user");
            try {
                System.out.println(user.getId());
                Cart cart=cartService.findCart(dressid,user.getId());
                if(cart!=null) {
                    //修改购物车数量
                    cart.setBooknum(bookSum+cart.getBooknum());
                    cartService.updateCart(cart);
                }else {
                    cart=new Cart();
                    cart.setBooknum(bookSum);
                    cart.setUser(user);
                    Dress dress=dressService.findDressById(dressid);
                    cart.setDress(dress);
                    cartService.addCart(cart);
                }
                map.put("msg", "true");
            }catch(Exception e) {
                map.put("msg", "false");
                e.printStackTrace();
            }
            return map;
        }
        //查找购物车
        @RequestMapping("/findCart")
        public ModelAndView findCart(HttpSession session) {
            User user=(User)session.getAttribute("user");
            List<Cart> cartList=cartService.findCartByUserId(user.getId());
            double sum=0.0;
            for(Cart cart:cartList)   {
                    sum+=cart.getBooknum()*cart.getDress().getPrice();
            }
            ModelAndView mv=new ModelAndView();
            mv.addObject("cartlist",cartList);
            mv.addObject("sum", sum);
            mv.addObject("state", false);
            mv.setViewName("cart");
            return mv;
        }
        // 单击"去结算"后获取要结算的Cart的集合,获取总金额,再跳转到dobuy.jsp
        @RequestMapping("/cartToOrder")
        public ModelAndView cartToOrder(HttpSession session,int ids[]) {
            User user=(User)session.getAttribute("user");
            List<Cart> cartlist=new ArrayList<Cart>();
            Cart cart=new Cart();
            double sum=0.0;
            for(int i=0;i<ids.length;i++) {
```

```java
            cart= cartService.findCartByCartId(ids[i]);
            cartlist.add(cart);
            sum+=cart.getBooknum()*cart.getDress().getPrice();
        }
        System.out.println(sum);
        ModelAndView mv=new ModelAndView();
        session.setAttribute("cartlist", cartlist);//将购物车存入session对象
        session.setAttribute("sum", sum); //将总金额存入session对象
        mv.setViewName("dobuy");
        return mv;
    }
    //修改购物车
    @RequestMapping("/updateCart")
    public void updateCart(HttpSession session,int cartid,int bookSum) {
        User user=(User)session.getAttribute("user");
        Cart cart=new Cart();
        cart.setId(cartid);
        cart.setBooknum(bookSum);
        cartService.updateCart(cart);
    }
    //删除购物车
    @RequestMapping("/removeCart")
    @ResponseBody
    public Object removeCart(HttpSession session,int cartid) {
        User user=(User)session.getAttribute("user");
        Map<String,Object> map=new HashMap<String,Object>();
        try {
            cartService.removeCart(cartid);
            map.put("state", "true");
        }catch(Exception e) {
            e.printStackTrace();
            map.put("state", "false");
        }
        return map;
    }
    //批量删除多个购物车
    @RequestMapping("/removeCartAll")
    @ResponseBody
    public Object removeCartAll(HttpSession session,int[] id) {
        User user=(User)session.getAttribute("user");
        Map<String,Object> map=new HashMap<String,Object>();
        try {
            for(int i=0;i<id.length;i++) {
                cartService.removeCart(id[i]);
            }
            map.put("state", "true");
        }catch(Exception e) {
            e.printStackTrace();
            map.put("state", "false");
        }
        return map;
    }
}
```

（7）navbar.jsp 的购物车超链接 onclick 事件绑定了 JS 函数 cart()，代码如下：

```
<a class="nav navbar-text" id="cart" href="javascript:void(0)" onclick="return cart()">
购物车<span class="glyphicon glyphicon-shopping-cart"></span> </a>
```

在 navbar.jsp 底部添加 cart 函数，代码如下：

```
<script type="text/javascript">
        //购物车
            function cart(){
                var username = $("#username").html();
                if(username == null || username == ""){
                    alert("请先登录!");
                    return false;
                }else{
                    location.href="findCart";
                }
            }
</script>
```

（8）在 dress.jsp 的购物车按钮中添加 JS 代码，代码在 dress.js 文件中。再在 list.jsp 中添加购物车按钮，并绑定下列 JS 代码：

```
// 加入购物车按钮
$(".add").click(function(){
    var username = $("#username").html();
    if(username == null || username == ""){
        alert("请先登录!");
        return false;
    }else{
        var dressid = $(this).parent().parent().children("a").attr("href").split("id=")[1];
        $.ajax({
            type:"post",
            url:"addCart",
            data:"dressid=" + dressid +"&bookSum=1",
            success:function(t){
                if(t.msg=="true"){
                    alert("添加购物车成功! ");
                }else{
                    alert("添加购物车失败");
                }
            },
            error:function(){
                alert("发生异常");
            }
        });
    }
});
```

（9）在 WebContent 下新建 cart.jsp 页面，具体内容请参见本书配套资源。

（10）运行与测试，效果如图 14.14 所示，至此购物车模块完成。

第 14 章　SSM 项目实战

图 14.14　购物车页面

14.7　订单处理与模拟结算模块设计

在图 14.14 所示的购物车中选择要结算的商品。单击"去结算"按钮，进入图 14.15 所示的页面 dobuy.jsp，进行订单确认。

图 14.15　订单确认页面

可以用默认地址，也可用新地址。确认无误后单击"提交订单"，新订单写入数据库，会进入支付界面 pay.jsp，如图 14.16 所示。这里不做真实支付，仅进行模拟支付，将数据库订单表中的支付状态字段改为 1 即可。

单击"模拟支付"按钮，进入图 14.17 所示的 paysuccess.jsp 页面，5 秒后自动跳回首页。

297

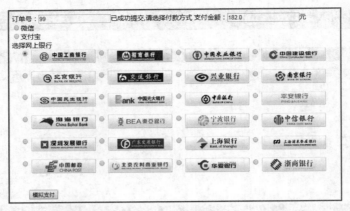

图 14.16　支付页面

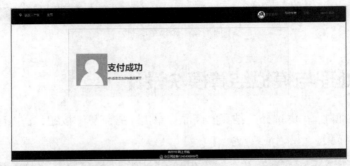

图 14.17　支付成功页面

实现步骤：

（1）在数据库中新建表 orders 用于存储订单数据，表 orderitem 存储订单明细数据，分别如图 14.18 和图 14.19 所示。

Alter Table 'orders' in 'dressdb'

Field Name	Datatype	Len	Default	PK?	Not Null?	Unsigned?	Auto Incr?
id	int	100		✓	✓		✓
money	double						
receiverAddress	varchar	255					
receiverName	varchar	20					
receiverPhone	varchar	20					
paystate	int	11					
ordertime	timestamp		CURRENT_TIMESTAMP		✓		
userid	int	11					

图 14.18　orders 表结构

Alter Table 'orderitem' in 'dressdb'

Field Name	Datatype	Len	Default	PK?	Not Null?	Unsigned?	Auto Incr?
orderid	int	100			✓		
dressid	varchar	100			✓		
buynum	int	11					
id	int	11		✓	✓		✓
*							

图 14.19　orderitem 表结构

包 com.lifeng.entity 下新建实体类 Order、OrderItem。

Order.java 代码如下：

```
public class Order implements Serializable{
private int id;
```

```java
    private double money;
    private String receiverAddress;
    private String receiverName;
    private String receiverPhone;
    private int paystate;
    private Date ordertime;
    private User user;
    private Set<OrderItem> orderitems=new HashSet<OrderItem>();
    //省略 getter,setter 方法
}
```

OrderItem.java 代码如下:

```java
public class OrderItem implements Serializable{
    private int id;
    private Order order;
    private Dress dress;
    private int buynum;    //省略 getter,setter 方法
}
```

(2) 在 com.lifeng.dao 下新建接口 OrdersDao,代码如下:

```java
public interface OrdersDao{
    // 改变订单状态-管理员
    public void updateOrder(Order order);
    // 新建订单
    public void addOrder(Order order);
    // 新建订单明细项
    public void addOrderItem(OrderItem orderItem);
    //订单 id 寻找
    public Order findOrderById(int id);
}
```

(3) 在 com.lifeng.dao 下新建接口 OrdersMapper.xml 映射文件,代码如下:

```xml
<?xml version="1.0" encoding="utf-8" ?>
<!DOCTYPE mapper PUBLIC "-//mybatis.org//DTD Mapper 3.0//EN"
"http://mybatis.org/dtd/mybatis-3-mapper.dtd">
<mapper namespace="com.lifeng.dao.OrdersDao">
    <select id="findOrderById" parameterType="int" resultType="Order">
        select * from orders where id=#{0}
    </select>
    <insert id="addOrder" parameterType="Order" keyProperty="id" useGeneratedKeys="true">
        INSERT INTO orders(money, receiverAddress, receiverName,receiverPhone, paystate,ordertime,userid)
        VALUES
            (#{money}, #{receiverAddress}, #{receiverName}, #{receiverPhone}, #{paystate}, #{ordertime}, #{user.id})
    </insert>
    <insert id="addOrderItem" parameterType="OrderItem" >
        INSERT INTO orderitem(orderid, dressid, buynum)
        VALUES
            (#{order.id}, #{dress.dressid}, #{buynum})
    </insert>
    <update id="updateOrder" parameterType="Order">
        UPDATE orders SET paystate=#{paystate} WHERE id= #{id}
    </update>
</mapper>
```

修改 mybatis-config.xml 文件，代码如下：
```xml
<configuration>
    <typeAliases>
        <typeAlias alias="User" type="com.lifeng.entity.User"/>
        <typeAlias alias="Dress" type="com.lifeng.entity.Dress"/>
        <typeAlias alias="Comment" type="com.lifeng.entity.Comment"/>
        <typeAlias alias="Cart" type="com.lifeng.entity.Cart"/>
        <typeAlias alias="Order" type="com.lifeng.entity.Order"/>
        <typeAlias alias="OrderItem" type="com.lifeng.entity.OrderItem"/>
    </typeAliases>
    <mappers>
        <mapper resource="com/lifeng/dao/UserMapper.xml"/>
        <mapper resource="com/lifeng/dao/DressMapper.xml"/>
        <mapper resource="com/lifeng/dao/CommentMapper.xml"/>
        <mapper resource="com/lifeng/dao/CartMapper.xml"/>
        <mapper resource="com/lifeng/dao/OrdersMapper.xml"/>
    </mappers>
</configuration>
```

（4）修改 DressDao 接口，添加修改商品方法 updateDress，代码如下：
```java
public void updateDress(Dress dress);
```

修改 DressMapper.xml，添加相应的 SQL 语句：
```xml
<update id="updateDress" parameterType="Dress">
     UPDATE dress SET quantity=#{quantity},sales=#{sales} WHERE dressid= #{dressid}
</update>
```

修改 DressService 接口，DressServiceImpl 实现类，添加 updateDress 方法如下：
```java
@Override
public void updateDress(Dress dress) {
    dressDao.updateDress(dress);
}
```

（5）在包 com.lifeng.service 中新建 OrdersService 接口，代码如下：
```java
public interface OrdersService{
// 改变订单状态
public void updateOrder(Order order);
// 新建订单
public void addOrder(Order order);
// 新建订单项
public void addOrderItem(OrderItem orderItem);
//订单 id 寻找
public Order findOrderById(int id);
}
```

（6）在包 com.lifeng.service 中新建 OrdersService 接口的实现类 OrdersServiceImpl，代码如下：
```java
@Service("ordersService")
public class OrdersServiceImpl implements OrdersService{
    @Autowired
    private OrdersDao ordersDao;
    //省略 getter,setter 方法
    @Override
    public void updateOrder(Order order) {
        ordersDao.updateOrder(order);
    }
    @Override
```

```
        public void addOrder(Order order) {
            ordersDao.addOrder(order);
        }
        @Override
        public void addOrderItem(OrderItem orderItem) {
            ordersDao.addOrderItem(orderItem);
        }
        @Override
        public Order findOrderById(int id) {
            return ordersDao.findOrderById(id);
        }
    }
```

（7）在包 com.lifeng.controller 下新建 OrderController.java，代码如下：

```
@Controller
public class OrderController{
    @Autowired
    private OrdersService orderService;
    @Autowired
    private CartService cartService;
    @Autowired
    private DressService dressService;
    //省略 getter,setter 方法

    @RequestMapping("/addOrder")   //添加订单
    public ModelAndView addOrder(HttpSession session,String addressType, String province,String city,
            String area,String detailed) throws Exception{
        User user=(User) session.getAttribute("user");
        double sum=(double)session.getAttribute("sum");
        String receiverAddress="";
        String telephone="";
        if(addressType=="newaddress") {
            receiverAddress=province+city+area+detailed;
        }else {
            receiverAddress=user.getShippingAddress();
            telephone=user.getTelephone();
        }
        List<Cart> list=(List<Cart>)session.getAttribute("cartlist");
        Order order=new Order();
        order.setMoney(sum);
        order.setPaystate(0);
        order.setReceiverAddress(receiverAddress);
        order.setReceiverName(user.getUsername());
        order.setReceiverPhone(telephone);
        order.setUser(user);
        order.setOrdertime(new Date());
        Dress dress=new Dress();
        for(Cart cart:list) {
            OrderItem oi=new OrderItem();
            dress=cart.getDress();
            //减掉库存
            if(dress.getQuantity()-cart.getBooknum()<0) {
                throw new Exception("订购数量不能超过库存!");
            }
            //扣减库存
            dress.setQuantity(dress.getQuantity()-cart.getBooknum());
```

```
                    //增加销量
                    dress.setSales(dress.getSales()+cart.getBooknum());
                    dressService.updateDress(dress);
                    oi.setDress(dress);
                    oi.setBuynum(cart.getBooknum());
                    oi.setOrder(order);
                    order.getOrderitems().add(oi);
                    //已下订单的商品从购物车中清理掉
                    cartService.removeCart(cart.getId());
        }
        orderService.addOrder(order);
        int orderid=order.getId();
        ModelAndView mv=new ModelAndView();
        mv.addObject("orderid",orderid);
        mv.setViewName("pay");
        return mv;
    }
}
```

（8）在 com.lifeng.controller 下创建 PayController，代码如下：

```
@Controller
public class PayController {
    @Autowired
    private OrdersService orderService;
    //省略 getter,setter 方法
    @RequestMapping("/pay")
    public ModelAndView pay(int orderid) throws IOException {
        Order order=orderService.findOrderById(orderid);
        order.setPaystate(1);//修改订单的付款状态为1,表示已付款
        orderService.updateOrder(order);
        ModelAndView mv=new ModelAndView();
        mv.setViewName("paysuccess");
        return mv;
    }
}
```

（9）在 WebContent 下新建 dobuy.jsp，代码参考源文件。
（10）新建支付成功页面 paysuccess.jsp，代码参考源文件。
（11）运行与测试。至此购物功能模块完成。

上机练习

读者在该项目的基础上继续完善其他功能，如用户注册、顾客后台管理、修改订单、卖家后台管理等。

思考题

1. 分页的实现流程是什么？
2. 简述购物车的实现步骤。

参考文献

[1] 黑马程序员. Java EE 企业级应用开发教程(Spring+Spring MVC+MyBatis)[M]. 北京：清华大学出版社，2017.

[2] 刘增辉. MyBatis 从入门到精通[M]. 北京：电子工业出版社，2017.

[3] 杨开振，周吉文，梁华辉，等. Java EE 互联网轻量级框架整合开发 SSM 框架（Spring MVC+Spring+MyBatis）和 Redis 实现[M]. 北京：电子工业出版社，2017.

[4] 杨开振. 深入浅出 MyBatis 技术原理与实战[M]. 北京：电子工业出版社，2016.

[5] 疯狂软件. Spring+MyBatis 企业应用实战[M]. 北京：清华大学出版社，2017.

[6] 朱要光. Spring MVC+MyBatis 开发从入门到项目实战[M]. 北京：电子工业出版社，2018.

[7] 陈恒，楼偶俊，张立杰. Java EE 框架整合开发入门到实战：Spring+Spring MVC+MyBatis[M]. 北京：清华大学出版社，2018.

[8] 肖睿. SSM 企业级框架实战[M]. 北京：中国水利水电出版社，2017.

[9] 高洪岩. Java EE 核心框架实战[M]. 2 版. 北京：人民邮电出版社，2017.

[10] Build Kunrniawan，Paul Deck. Servlet、JSP 和 Spring MVC 初学指南[M]. 北京：人民邮电出版社，2016.